Cultural Anatomies of the Heart in Aristotle, Augustine, Aquinas, Calvin, and Harvey

Giotto di Bondone. Charity. Scrovegni Chapel, Padua. Courtesy of: ART Collection / Alamy Stock Photo

Marjorie O'Rourke Boyle

Cultural Anatomies of the Heart in Aristotle, Augustine, Aquinas, Calvin, and Harvey

palgrave
macmillan

Marjorie O'Rourke Boyle
Centre for Medieval Studies
University of Toronto
Toronto, ON, Canada

ISBN 978-3-030-06699-4 ISBN 978-3-319-93653-6 (eBook)
https://doi.org/10.1007/978-3-319-93653-6

Cover image: Andreas Vesalius, *De humani corporis fabrica libri septem*, 1555, p. 705. This image is from Vesalius's personal annotated copy. Reproduced by permission of the owner, Dr. Gerry Vogrincic. Photo Courtesy of the Thomas Fisher Rare Book Library, University of Toronto.

This Palgrave Macmillan imprint is published by the registered company Springer Nature Switzerland AG
The registered company address is: Gewerbestrasse 11, 6330 Cham, Switzerland

The new chapters in this book are dedicated to scholarly friends:
on Aristotle, to Vivian Nutton; on Augustine to †Walter H. Principe;
on Aquinas, to David Novak; on Calvin, to †William J. Bouwsma.

Preface

The heart is familiar literally as the physical organ for the vital circulation of the blood and metaphorically as the profound site of emotions, iconic ♥ for love. But it was not always so. This book revises ordinary assumptions by original research on significant concepts of the heart in important ancient to early modern thinkers. For, the identity and the function of their hearts differed significantly from modern physiological knowledge and psychological attribution. Their hearts were the agents for or against the divine law; they were the seats of the soul with its intellect and imagination; and they were the furnaces of the blood. *Cultural Anatomies of the Heart in Aristotle, Augustine, Aquinas, Calvin, and Harvey* discovers the unfamiliar heart in four new substantial and surprising studies. These investigations establish the medical foundation of the heart in Aristotle's natural philosophy; reveal Augustine's social critique of his outlaw heart; analyze grammatically and logically the heart of Aquinas's natural law of morality; and present theologically Calvin's personal emblem of the divided heart in hand. The book also reprints one of the author's eleven articles on the heart published in scholarly journals. This article unmasks William Harvey's heroic persona as the discoverer of the heart's circulation of the blood, a mystery physicians believed "almost known to God alone."

These cultural anatomies are innovative in their scholarly topics and interdisciplinary methods. Anatomy was historically the physical dissection of the body for medical knowledge. It was practiced traditionally on an animal carcass, eventually also on a human cadaver, usually that of a criminal. As a cultural anatomy of beliefs and ideas about the heart, this book is a scholarly examination of texts for conceptual knowledge. It does not palpate the surface of the heart but cuts into its substance with acute incisions to lay bare human existence for inspection and analysis. For context and continuity, this Preface to the book integrates the author's published research on the heart with the four new anatomies here. That research spanned a history from an ancient poem of triumph in the wilderness to an early modern scientific record in a laboratory. It began by plunging into the biblical "heart of the sea" to fathom the divine punishment that made a path for the Exodus of the Israelites from Egyptian slavery. And it ended by probing beneath the skin of a ligated arm, through its labyrinthine venous membranes, to a reasoned inference of the blood flow originating in the heart. The hearts of Aristotle, Augustine, Aquinas, and Calvin anatomized in this volume advance the history of the heart significantly.

Ancient Realities

In the beginning, in the Hebrew Bible, *lēb/lēbab* translated as "heart" appeared more than 800 times as its principal anthropological concept. Its initial entry in the standard Hebrew lexicon is "the quivering, pumping organ, the heart." However, since not even Harvey's authoritative *Exercitatio anatomica de motu cordis et sanguinis in animalibus* (1628) modeled a pump for the heart's circulation of the blood, that definition is a serious anachronism. Christian adoption of the Hebrew Bible as the Old Testament extended the importance of biblical heart through citation and interpretation. However, for want of both linguistic knowledge and historical method, through classical acculturation Christian exegesis early subsumed Hebrew heart to Greek philosophy and medicine and usually in Latin translation. Those applied classical disciplines so overwhelmed the biblical sense that modern reference works impute anachronistic and foreign meanings to biblical heart. Christian exegesis has insisted that the Bible meant by "heart" a classical facultative psychology of the intellect, with reason and will. It has further asserted that the Bible had a precocious and precise medical knowledge of the heart. A major biblical dictionary retroactively assumes Harvey's heart as the circulator of the blood to be the ancient basis for "heart" as the seat of personal energies, from which circulate all mental and moral activities. This is obviously another anachronism. The physical heart's circulation of the blood, which was only discovered in the seventeenth century C.E., and that with an Aristotelian philosophical inference, cannot be the premise for ancient biblical "heart" as the human soul or mind. Yet, standard modern reference works erroneously claim for the Bible a precise physical cardiac knowledge. This author's publications on biblical "heart" in the Torah, prophets, histories, and psalms have exposed the context of their "heart" as legal, not psychological or physical.

"'In the Heart of the Sea': Fathoming the Exodus" interpreted "heart" as the agent of the law in the most ancient biblical text, the Song of the Sea (Exodus 15) in praise of the divine deliverance of the Israelites from Egyptian enslavement. For Judaism, that Exodus was the normative historical event and the traditional basis for faith. As Yhwh (God) acted in that poem, his nostrils blew the waters upward into a heap and "the deeps congealed in the heart of the sea" (v. 8). Modern exegesis has assumed that the heart of the sea was its interior midst on the false analogy of the physical heart situated in the middle of the human body. Yet, the several other biblical phrases "the heart of the sea(s)" all referred to Yhwh's punishment by death of Israel's enemy. This research identified "heart" as the active receptor or rejector of divine law. Heart searched, inclined, and moved toward—or against—the observance of God's commands. Its analogue was not the interior physical heart, for there was no such anatomical knowledge among the ancient Israelites, but the exterior physical legs with feet, which Israelite physicians did know and treat. As definitive of human character, biblical heart walked upright in God's ways or it departed from them by stumbling, backsliding, falling, breaking. Heart was a concept for lawful observance, not for psychological will, yet another anachronism, as in Augustine's psychology. Biblical heart as movement on a particular way coincided in Exodus 15:8 with the trajectory of a storm at sea, where the waves rose up to overwhelm and drown the pharaoh's charioteers in pursuit of the fleeing Israelites. Its phrase "in the heart of the sea" meant in the judgmental movement of the sea, not in its locational midst. The drowning of Israel's enemy was a legal process of divine judgment by talion—like for like—that weighed down the charioteers of the hard-hearted pharaoh who sank like stones in the hardened waves of the sea.

"The Law of the Heart: The Death of a Fool (1 Samuel 25)" rejected the standard proof text for a biblical cardiology, the story of Nabal whose "heart died within him, and he became as a stone" (v. 37). Although his fictional death has been diagnosed medically, there is no evidence for the assumed Egyptian source of a biblical cardiac knowledge. Nabal's death is explained contextually by Hebrew philology and biblical law. The story of Nabal played on his name, *nābāl* "fool," which denoted serious sin. A biblical fool committed extremely disorderly and unruly acts that endangered or destroyed social relationships. Nabal's refusal of David's entitlement to a share of his harvest as a tariff for protection committed a breach of customary law (*nĕbālâ*), a grievous anti-social sin. Nabal's sin was compounded by his further transgressions of the laws of practicing hospitality, of feeding an emancipated slave, and of welcoming a sojourner. His wife Abigail intervened to rescue his household from David's retributive justice, calling her husband's unlawful behavior by the rare term *nĕbālâ*, "extinction." It was precisely when Abigail reported her diplomatic success that Nabal's heart failed because of his hardened—metaphorically stony—defiance of those laws. The biblical story was about Nabal's wanton transgressions, not his cardiac physiology.

The New Testament cited the Hebrew Bible on "heart," most importantly in Jesus's command to observe its law of loving God "with your whole heart" (Deuteronomy 6:5; Matthew 22:37). It repeated its concepts of the divine commandments and covenant as written on the heart (Jeremiah 31:33; Hebrews 8:10). It employed idioms of saying or thinking "in the heart" about observance of the divine law, and it condemned the hardened hearts of fools who disobeyed it. However, the Christian gospels also altered strict Judaic observance of ritual and judicial precepts, and converts to the Christian faith included non-Jews, or Gentiles of other nations and cultures. Christian interpreters of Hebrew biblical "heart" assumed that its *lēb/lēbab* meant the physical organ of Greek medicine and philosophy, although it did not. They imported extraneously those classical disciplines as sources for incorporating Hebrew biblical "heart" with a Christian heart. An incident in Luke's gospel, not about a dead metaphorical "heart" like Nabal's but about a live physical heart like Christians', illustrated the cultural assimilation. In that gospel, upon the discovery that the dead body of the crucified Jesus was missing from its burial, two of his disciples conversed about the astonishing news of his empty tomb with an encountered stranger, who interpreted it scripturally then vanished. "They said to each other, 'Did not our hearts burn within us while he talked to us on the road, while he opened to us the scriptures?'" (Luke 24:32). Their burning hearts did not essentially denote an ardent affective heart, although that psychological connotation became attached to the material heart as the seat of the immaterial soul. A burning heart was a belief of the Greek physiology that defined the vital principle of animal bodies as heat. (A cold animal was a dead animal.) That quality of heat devolved to natural philosophy ("physics"), which classified the qualities of the physical elements—fire, air, earth, water—as hot, cold, dry, or wet, whether solely or complexly. Luke's precise knowledge in the first century C.E. is indeterminable. The principal texts for the heart as the source of bodily heat were Aristotle's animal books of the fourth century B.C.E. and an anonymous *De corde* tentatively of the third. That latter text circulated among the "Hippocratic" writings but, save for fragments in citation, its manuscripts were lost until the medieval recovery of Arabic or Greek copies. Although Luke the evangelist was called a physician in the tradition of the Epistle to the Colossians 4:14, there is no corroboration of that profession, and the authorship of that biblical letter is disputed. Luke's burning hearts meant live hearts. The burning hearts of those Christian disciples embodied their belief that Jesus, whom they had encountered on the road to discovery as the stranger, was no longer buried as dead but risen as alive.

"Broken Hearts: The Violation of Biblical Law" studied the imposition on the Bible of yet another anachronism, brokenhearted as affective for "grief or disappointment." It exposed the exclusion in Luke's gospel of the prophet Isaiah's mission "to bind up the broken hearted" from Jesus's reading of it in the synagogue to announce his own mission (Isaiah 61:1; Luke 4:16–21). This research discredited medical diagnoses of the Hebrew biblical verses about broken hearts since ancient medicine rarely ventured into cardiac pathology and it did not bandage hearts. There were, however, therapies for dislocated and fractured bones. The Hebrew Bible and the Christian Old Testament both paralleled broken bones with disabled hearts. Broken hearts were lawbreaking hearts. They were embodied by broken or crippled legs, which disabled an upright movement toward observance of the divine law. The study documented how Luke's gospel incorporated the biblical metaphor of walking—or not—in the divine law and also the classical culture of erect bipedality to characterize Jesus as upright in observance of that law. It then documented the mistaken insertion of Luke's omitted verse about Isaiah's "brokenhearted" (61:1) into twelve centuries of gospel manuscripts and editions. It further reported medieval and later translations of *brokenhearted* as "contrite," a reversal of the biblical broken "heart" as lawless and not repentant. The mistranslation seriously affected the theological controversy of the Protestant Reformation about the nullity of the will toward grace, Martin Luther's principal doctrine. This investigation finally identified the origins of the English definition of *brokenhearted* as "grieved," not in biblical but in secular literature, as in Chaucer and Shakespeare where it meant to "break forth" or "crack up" in lament.

The "heart" (*lēb/lēbab*) in the Hebrew Bible, meaning the human agent of the divine law, became foundational for Christian thinkers who later interpreted that word as the physical heart, the seat of the soul. That secular foundation for the heart was the natural philosophy of Aristotle (384–322 C.E.) in his books on animals. They were unaccountably lost until their recovery in Arabic then Greek copies, which became fundamental doctrine in medieval universities. Aristotle identified heat as the vital principle of animals, and he distinguished blooded animals, which originated from the fetal heart, as the noblest living beings, culminating in human beings. Aristotle's heart was not just a bit of biology but the seat of the human soul with its mental faculties and operations. Thus philosophical and theological deliberations about the human soul—an essential topic—were in the magisterial Aristotelian tradition necessarily cardiocentric. Aristotle believed that his research by manual dissection and his reasoning by causal inference happened by cardiac movement. The heart acted to know itself, for it empowered and controlled the mind in the body. This book of new cultural anatomies begins with interdisciplinary research for a revisionist evaluation of Aristotle's cardiac physiology. It is newly set in the contexts of his philosophy for rational principles and of his rhetoric for empirical models. Both are set in the context of the social usage of ordinary artifacts.

"Aristotle's Cardiac Vessel" (new) does not begin scientifically, as in the academic history of medicine, with Aristotle's observation of a cardiac blood spot in the embryo of a dissected chick. It begins culturally with his unexamined cardiac models, a vase and an oven. It contextualizes Aristotle's writings on the heart by arguing from his mindset to his cardiology. As a natural philosopher, Aristotle sought knowledge of the causes of animal movements because an innate principle of movement distinguished nature from artifact. Investigators of nature, as in the Hippocratic medical writings, frequently resorted for explanations to metaphors, such as the heart as a flaming hearth with the lungs as airy bellows to cool it lest it combust. Aristotle's cardiology prominently employed artifacts to understand and to describe nature. Although he dissected animals, he did not vivisect them, thus he had no empirical observation of a functioning live heart. He had to infer its physiology from a physics of observable universal movements and by a metaphysics of a reasoned final cause.

This study offers a new premise for the history of the heart. It challenges the criticism that Aristotle's cardiology failed because his definitive model, a "vessel," was structural not functional. It researches contrarily that the shape of Greek vessels was in fact determined by function. It documents Aristotle's knowledge of their craft and technology, his use of vessels in his animal research, and his comparison of household pottery with bodily vessels for fluids, particularly the blood. It explicates Aristotle's basic distinction in his natural philosophy between containers and contents. For, his prime philosophical example for movement, defined as a change of place, was the amphora of wine. It was a temporary storage for that liquid to be poured out elsewhere, from one place to another place. This study identifies the common pointed amphora of wine as Aristotle's dynamic, not static, heuristic for the heart as a vessel whose content of blood was regularly displaced from that temporary container into another place. It then addresses how Aristotle intuited that the cardiac vessel discharged its supply of blood into the tributary blood vessels for distribution in the body.

This study then explains how Aristotle developed his notion of the heart as the bodily hearth of vital heat by appropriating a vessel auxiliary to the amphora, the oven. It explains the hot oven as his heuristic for how the innate heat of the heart expands the blood in the cardiac chambers: by pulsation through their sinews upward through the aorta and finally into the attached blood vessels for distribution throughout the body. It offers how, analogous with the stopper of an amphora or an oven, he inferred that death was cardiac failure caused by the closure of the juncture between the heart and the great blood vessel and/or the aorta. Aristotle's reason and research exploited the ordinary to investigate the extraordinary, the hidden natural hearts of animals from visible manufactured vessels. Contemporary physicians would have recognized his model of a cardiac vessel from their own use of an amphora of wine because they dispensed it extensively to their patients as therapies. Aristotle's heart as a "vessel" was obviously functional and viably heuristic. His cardiac physiology was far more advanced than has been acknowledged, for the diligence of his research and the ingenuity of his reasoning.

The loss Aristotle's animal books until their medieval recovery created a gap during which Christian thinkers interpreted the heart by their available sources. That could include an oral conversation with a physician, such as Augustine's with Vindicianus, who by chance crowned him the winner of a poetry contest. The notable medical source was the writings of Galen of Pergamum, a Roman physician and philosopher in the second century C.E. who argued for the brain, not the heart, as the seat of the human rational soul. His prolific writings long informed medical practice. Still in the seventeenth century, Harvey, as the anatomist for the College of Physicians, London, was accused by its fellows that his discovery of the heart's circulation of the blood breached their professional oath to uphold Galen's medicine.

Medieval Thoughts

The most influential patristic theologian, Augustine (354–430 C.E.), examined his heart profoundly, earning the iconographic attribute of the burning heart. Conversation with the friendly physician Vindicianus introduced him to the cardiac theory of Herophilus, an Alexandrian physician in the third century B.C.E., who dissected and even vivisected. "Augustine's Heartbeat: From Time to Eternity" took the pulse of that theologian as he meditated toward eternity by transcending time with his whole heartbeat. His singular contribution on the heart was the climax of his *Confessions* in a conversation at Ostia with his mother. It progressed from their sensory perception of an earthly garden to their mental attainment of the eternal paradise. Its climax has been canonized as an ecstatic revelation of God inspired by Platonist metaphysics. However, that interpretation has ignored Augustine's heart and mistranslated his stated achievement, *toto ictu cordis*, which means "with a whole heartbeat." This study argued that Augustine's experience was a rational meditation from time to eternity by conflating the beats of meter, music, and medicine. He appropriated the term *ictus* "beat" from his practice as a grammarian to mark the metrical stresses of words. Augustine's professional experience coincided with his philosophical conviction of the power of musical rhythms to attain infinity through numeration. Decisive for his meditation was the medical pulse lore of Herophilus, who applied meter and music to cardiology. Herophilus posited that the heartbeat made music by four distinct rhythms, from infant to elder. Augustine believed the beautiful and perfect beat was his youthful trochee ¯˘. His method from time to eternity "with a whole heartbeat" scanned onomatopoetically as a triple trochee, *toto ictu cordis* ¯˘ ¯˘ ¯˘. Augustine conversed with his mother step-by-step by meditating on creatures then transcending their temporality by tapping "with a whole heartbeat" the absolute trochee, the eternal Word, *Verbum* ¯˘, their Creator. Augustine's ingenious notion of the heartbeat making music unto God was a confident rational exercise premised on classical harmony, from bodily rhythms to cosmic rhythms, to their creative impulse. God was "the pulser" of "the ears" of Augustine's heart, its auricles as the dwellings of his mind and spirit.

The association of biblical "heart" with observance—or not—of the divine law was so frequent that Christian theologians could not have ignored those verses. However, they interpreted that heart philosophically. Two Christian swerves from the biblical law of the heart were Augustine's social law of the heart and Aquinas's natural law of the heart. They not only proposed new interpretations but also practiced distinct methods, the rhetorical or the philosophical. This book anatomizes those hearts culturally.

Augustine's adolescent theft of pears is a notorious yet perplexing episode of his *Confessions*, personal epideictic rhetoric in praise of God and blame of self. Although his theft has been discussed philosophically and theologically, **"Augustine's Law of the Heart: Thieves' Honor" (new)** researches it historically, rather than theoretically. As Augustine pondered his theft, "Thy law, o Lord, certainly punishes theft, and the law written in human hearts, which even injustice itself does not delete, for indeed what thief suffers a thief with an even mind." This new research documents and interprets Augustine's law of the heart as conformity to human custom, in distinction to the biblical law of the heart as obedience to divine command. It studies his indictment of himself and his companions for their wanton theft of pears in its Roman socio-cultural contexts, especially the *ludus*, the indulgent adolescent period of sexual "play." It examines Augustine's blame of their theft on Roman exemplary rhetoric, which promoted for human imitation the god Jove's rapes euphemized as thefts. It identifies the comedic rapist Chaerea, the adolescent protagonist of Terence's *Eunuchus*, as Augustine's model to incite the applause of his peers for their collaborative theft. It explains their complicity in Augustine's law of the heart as social affinity, which he cited as proverbial thieves' honor. The proverb amplified how Augustine's phrase "law written in human hearts" acted as a law unto itself in an anarchic sense. This study researches that, under Roman law, honor among thieves, whether in banditry or piracy, was outlawry. It acted not only outside divine law but also outside human law. Outlawry operated anthropologically by a code of honor and shame, as a social, not a moral, concept that determined precedence in a group. Augustine confessed that his lust, impelled by the custom of adolescent sexual play, exceeded the civilizing boundaries of friendship and marriage. As customary behavior, Augustine's law of the heart required a boastful reputation among his peers for his sexual play—or else face ostracism or retaliation from his group. His law of the heart was not about lawful obedience or disobedience, as was the biblical law of the heart, but about social conformity or nonconformity. Augustine's law of the heart was customary behavior from a proverbial social sympathy: "like likes like" and so collaborates in deed.

Aristotle's lost records and lectures of natural philosophy on the heart were recovered and finally translated into Latin by Michael Scot from an Arabic version. That translation based the scholastic Albert the Great's (c. 1200–80) *De animalibus* (On animals), which broadly introduced Aristotle's heart to Latin readers in the thirteenth century. With an Aristotelian resurgence, scholars in universities energetically deliberated the role of the heart as a repository for philosophical and theological ideas about the soul and its faculties and operations. For the distinctly human soul, the heart was principally the seat of reason. Yet, for scholastics intent on promoting good morality, the cardiac causes or consequences of the passions of the soul also became a major topic of psychology. Those passions were not emotions, the invention and neologism of a physician only in the late nineteenth century. The medieval passions denoted not a feeling but a state, an accidental happening (passivity) from a sensory impression on the soul. As scholastics deliberated, did the soul move the body or the body, the soul?

"The Wonder of the Heart: Albert the Great on the Origin of Philosophy" introduced Albert's astonishing idea that wonder happens by a cardiac systole, or contraction. This study developed Albert's reasoning from Aristotelian cardiocentrism, which seated the soul in the heart, in alliance with Christian writings on the passions of the soul. It researched Albert's comparative usage of the difficult terms of his definition of cardiac wonder in his paraphrase of Aristotle's *Metaphysica* and in his philosophical and biblical commentaries. It documented Albert's own cardiocentrism as a natural philosopher and related it to other medieval discussions of the functions of the heart for the passions of the soul. It treated the theory of the passions of the soul as cardiac movements in systole or diastole and the particular Christian notion that typified wonder as fear. It reported extensively on the particular attribution of fear to a cold cardiac contraction. It detailed Albert's reasoning that the soul is in the heart, a contraction of the heart causes cold passions, a type of cold passion is fear, a type of fear is wonder, therefore wonder is a passion caused by a contraction of the heart. It explicated Albert's theory that cardiac systole caused wonder because passive contraction in ignorance suspended active expansion in knowledge. Wonder, as a suspension in ignorance by a cardiac contraction, was a cold passivity like fear. The study reported how Albert integrated cardiac physiology not only with the medieval passions of the soul but also with the classical medical qualities and humors of the body. It explained how he defined philosophy as cardiac movement beyond wonder to knowledge by overcoming cold passivity in systole, contraction, with hot activity in expansion, diastole. For, he thought that diastole moved the hot blood and its spiritous substance from the heart for distribution in the body for optimal operation by the soul's powers. Albert's integrated psychology and physiology of the heart exemplified a medieval aspiration for knowledge of the heart and ultimately for knowledge of the cause of knowledge itself as the heart.

Albert's illustrious pupil Thomas Aquinas (1224/25–1270) embodied the scholastic ideal of causal knowledge until an extraordinary personal fear caused not the wonder originating philosophy but the stupefaction ending it. "Aquinas's Natural Heart" began by documenting Aquinas's writings on the irascible passion of fear and its type stupor, which experience caused him to repudiate all his writings as "chaffy." It explicated Aquinas's rejection of the cause of fear as a cardiac systole. For, his *Summa theologiae* and other writings asserted the primacy of the soul over the body. That primacy reassigned the cause of all passions to an alteration of the soul, which then accidentally affected the cardiac movements of systole and diastole. This research reported Aquinas's extensive mention of the heart in his biblical exegesis and philosophical treatises. He consulted Aristotle's heart in a Latin translation of his Greek animal books by William of Moerbeke, a fellow friar. This study exposed Aquinas's consistent rejection of the Aristotelian cardiocentrism dominant in medieval universities. Its tradition posited that the soul, as the vital principle of the body, existed in the body as domiciled in the heart, which was designated the primary bodily organ. This study treated particularly Aquinas's late brief *De motu cordis* (On the movement of the heart) about the relation of the heart to the passions of the soul. It analyzed that neglected cardiac treatise as a decisive instantiation of Aquinas's unique psychological doctrine, the soul as the form of the body. It revealed how Aquinas significantly revised the contemporary Aristotelian cardiocentrism. Aquinas argued that the soul did not reside in the heart but informed the body and thus cohered with every bodily organ. He still maintained with the Aristotelian tradition that the heart was indeed the natural mover of the body, even though the heart was moved by the soul. But he redefined the meaning of Aristotle's heart as "natural." Aristotle designated as "natural" the possession of an innate principle of movement. Aquinas distinguished the heart as "natural" only insofar as the soul informed the entire body—but not as resident in the distinct organic heart.

Aquinas's 103 texts had "heart," as Latin *cor*, more than six thousand times, from his first theological writing to his last philosophical treatise, *De motu cordis*. The vast majority of those hearts occurred in biblical quotation and exegesis, specifically in the commentaries he produced as a *magister in sacra pagina*, "master of the Bible," his academic authorization. Yet, for want of linguistic knowledge and historical method, his hearts could not adhere to their original biblical sense. Aquinas's ultimate extraction of the soul from the physical heart in *De motu cordis* nullified his prior discussions of "the law of the heart," both biblical and natural. However, it was his prior discussion of "the law of the heart" that has endured in modern ethics, especially in current political arguments about a natural moral law. **"Aquinas's Law of the Heart: Natural Reason" (new)** is a historical investigation independent of those contemporary discussions to understand his medieval meaning.

This new cultural anatomy publishes that Aquinas's *Summa theologiae* reconstructed Augustine's *Confessions* about "the law written in human hearts" as a proof text for his own natural law. Considering whether the natural law was impermanent, Aquinas reasoned, "But to the contrary is what Augustine says, 'Thy law has been written in human hearts, which not even injustice itself deletes.' But the law written in human hearts is the natural law. Therefore, the natural law cannot be deleted." Aquinas's citation both omitted and added words, reordering the syntax to substantiate by Augustine's authority his own theory of the natural law of the heart. This study begins by demonstrating how Aquinas's inadequate schooling in the trivium disadvantaged him to read Augustine's rhetoric accurately. It then presents Aquinas's philosophical dependence for his natural law on Aristotle's physics of movement and psychology of reason. It exposes the contradictions between the premises of an uncreated or a created universe, and between the principles of an intrinsic nature or an extrinsic law. It presents Aquinas's revision of Aristotle's definition of "natural" as possession of an intrinsic principle of movement to "natural" as participation in an external principle of movement, finally God. It analyzes the contradiction between those principles then Aquinas's proposed reconciliation of them by a metaphysics of participation.

It discovers Aquinas's dependence on an undetected source for his first precept of the natural law, "good is needing to be done and pursued, and evil is needing to be avoided." That source was Nemesius of Emesa's misattributed *De natura hominis* on the demonstrative topic of the rational nature. This study explains that, as a demonstrative topic in the Aristotelian tradition, Aquinas's first precept of the natural law was a premise for conclusions that were logically necessary but not ethically obligating. An analysis of Aquinas's consistent usage of the passive periphrastic construct to mean a command evidences a certain inconsistency with Latin grammar, where it only implied a logical conclusion. His first precept of the natural law was grammatically not a command but an indication, as consistent with his claim of its self-evidence. Aquinas did not treat how his natural law, as a premise for demonstrative reasoning and a self-evident indication, related to any obligation to act upon it or its logical conclusions. Although he posited that human reason, as divinely imprinted by the eternal law, commanded the human observance of the natural law, he did not discuss whether everything reasonable was therefore obligatory. Aquinas's theory evidenced culturally an education in the trivium deficient in rhetoric, for his misconstrual of Augustine's sentence, and in grammar, for his misconstrual of Latin commands. The practical context of his *Summa theologiae* as a reference for Dominican confessors may also account for some unclarity of his law of the heart as natural reason.

Early Modern Discoveries

In contrast to Aristotle's cardiocentrism, the medieval recovery of which texts inspired some scholastics, was Plato's reputed assignment of the noble rational soul to the brain and the lesser spirited soul to the heart. However, Plato did not apportion those faculties of the soul to those organs but only to the cranial and thoracic venters, which were occupied as well with other organs and structures. Marsilio Ficino (1433–1499), the Renaissance philosopher who broadly published Plato for Western thought, was a syncretist who integrated Plato's dialectic with biblical "heart." The study "Pure of Heart: From Ancient Rites to Renaissance Plato" explained Ficino's alliance of Jesus's beatitude for the "pure of heart" (Matthew 5:8) with Plato's perfectly clean bosom. This study discerned the oddity of Ficino's praise since Plato himself never advocated purity of heart. Further, the ancient sources never described Plato as clean but instead vilified him as a filthy polluter with his atheism and sophistry.

This research reported that comparative religious cleanliness very rarely involved the heart, and that the biblical phrase "pure of heart" is still undefined. It discussed ancient purification as essentially a separator, and it reported how purification particularly separated the sacred and the profane in Greek ritual sacrifice, which Plato's philosophy mirrored. It analyzed purification as a separator in Plato's logical science of division, where purification of the intellectual soul was the aim of his dialectic. By the rational discipline of refutation, his dialectic purified the soul by purging the evil of ignorance and retaining the good of wisdom. It analyzed Plato's purificatory division of love into different types to explain Ficino's praise of him as perfectly clean—not by moral purity but by philosophical integrity. It studied Ficino's usage of "bosom" to represent Plato as the expert statesman, a clean and pure wise separator from the ignorant crowd. It researched the biblical parallelisms of bosom and heart, and the biblical and ecclesiastical type of Abraham's bosom. It disclosed how Ficino defied the papal damnation of the pagan philosophers by gathering Plato into that topical Abraham's bosom in limbo. From there Jesus in his harrowing of hell rescued Plato's soul and carried it to the beatific vision of Plato's philosophy. Ficino did not baptize pure-hearted Plato as a Christian but adopted him as a Jew by an accord of natural and moral law for universal precepts and a correct life.

The Protestant Reformation in the sixteenth century tended to rupture the alliance of theology with philosophy. The reformer Jean Calvin (1509–1564) dismissed medieval scholasticism as brainy for its notional faith and demanded an affective Christian faith rooted in the heart. In continuity with his education in Renaissance humanism, which tended to subordinate logic to rhetoric, Calvin argued from Scripture rhetorically, and skillfully so. **"Jean Calvin, Heart in Hand" (new)** examines his novel integration of cardiology with theology. It reports his traditional exegesis of biblical "heart" by the physical heart of philosophy and medicine. It then interprets the iconography of Calvin's personal seal, which depicted in his hand a cloven heart, divided like all sinful hearts between hypocrisy and sincerity. Although he favored an Aristotelian heart as the seat of the soul, he condemned its faculties as disabled toward salvation ever since Adam's original sin of idolatry. Calvin severely characterized the heart as evil: viciously depraved, fraudulent and perverse, corrupt and stinking like the putrefaction of intestinal worms. For, against papist arguments for freedom of choice, he argued that the divided heart was unable to choose integrity and the hardened heart was unable to choose flexibility. This study indicates how Calvin complemented that biblical diction for "heart" with the medical terms for the physical cardiac movements of contraction and expansion. It explicates his teaching about the Spirit's extraction of the stony deformed heart and his implant of a soft healthy heart. It presents Calvin's doctrine of a fallen humanity in need of the Spirit's imprint of his own personal seal on their hearts to recreate and quicken them and to stabilize their faith. Criticizing scholasticism for a speculative faith at the tip of the brain, Calvin promoted an affective faith in the intimate heart for active transformational behavior. He did not ally that behavior with a natural law, however, for he believed Christian integrity originated in a persuasion of the conscientious heart, not in a proof of the rational brain. This study treats Calvin on the affects of the soul by cardiac physiology and humoral theory, notably his rejection of the Stoic therapy for fear. It discloses how Calvin's doctrine of the sanctification of the heart depended on Aristotle's definition of nature as the possession of intrinsic movement. Because Calvin believed that fallen human nature did not possess an intrinsic movement toward salvation, the Spirit intervened by his personal intrinsic movement to reverse to himself as good the natural movement of the heart as evil.

The final four studies presented valuable but unknown topics for the most important book in the history of medicine as a scientific discipline, William Harvey's (1578–1657) *Exercitatio anatomica de motu cordis et sanguinis in animalibus* (Anatomical exercise on the movement of the heart and the blood in animals). These studies began with overlooked and unexamined models for his cardiovascular investigations: a title page, a comedic soliloquy, a sluice gate, and a swear word. As the anatomist for the College of Physicians, London, he composed his book both defensively and aggressively when some of its fellows accused him of breach of oath with the medical tradition. Harvey took Aristotle as his master, from his texts but also from the demonstrations of Girolamo Fabrici d'Aquapendente, his Paduan professor. But Harvey then dissented. Three studies examined his ingenuity in applying to anatomy his formative education in Renaissance humanist practices of observation, comparison, and reason. Given Harvey's grammar schooling in the humanities and his university degree in the arts, these studies appropriately applied to his texts literary criticism. Another study discovered how Harvey's voyage to Venice from Padua while a medical student afforded him the technological model of the double-mitered sluice gate for the function of the venous membranes in the heart's circulation of the blood.

"William Harvey's Anatomy Book and Literary Culture" interpreted the attributed Benedictine motto *Ora et labora* on the title page of his *Exercitatio anatomica de motu cordis et sanguinis in animalibus.* This study described its depiction of a heraldic angel and detected its motto, "Pray and Work," as Harvey's announcement of anatomical "work." It researched the origin and usage of the attributed Benedictine motto and of the adopted motto of the chief of clan Ramsey. It documented Benedictine associations of Harvey's royal patrons and patients, James I (VI of Scotland) and Charles I of England, the latter to whom, as the nation's "heart," Harvey dedicated his book. It further detailed Harvey's educational associations with the Benedictine monasteries in Canterbury at its cathedral and at St. Augustine's Abbey; in Cambridge at Caius College, which was built from the ruins of Ramsey Abbey; and in Padua at the reformed Abbey of Santa Giustina, whose basilica housed the tomb of the evangelist Luke, who was still regarded as a physician. Harvey thus knew very well that Benedictine work was the collection, copying, and care of traditional manuscripts. Harvey's titular motto *Ora et labora* in his anatomical book challenged the medical profession. He criticized it for practicing, in his opinion, a secular version of Benedictine scholarly labor by transmitting its own traditional manuscripts as sacrosanct. His proem reported his provocation to publish his book by the slander of certain colleagues who accused him of a breach of his oath to perpetuate Galen's writings about the heart and about the blood. Harvey's book published his own ideal and practice of anatomical labor as research by observation, demonstration, and reason. It rejected the College's sworn reliance on books, although humanist editions of medical texts usefully provided him precedent knowledge, theory, and opinion. They served him at the very least as hypotheses for rejection. Harvey intended his book to participate in the literary republic by retaining the humanist ideal of a learned exchange but by reforming dependence on previous books by promoting his own anatomy book.

"William Harvey's Soliloquy to the College of Physicians: Reprising Terence's Plot" developed Harvey's subversion of Renaissance humanism by redefining its literary republic from replicating traditional manuscripts to publishing his novel texts. Harvey characterized himself as Demea in Terence's comedy *Adelphi* (The brothers) to persuade his detractors to be open-minded. Schoolboys in the seventeenth century still learned colloquial Latin from Terence's comedies. Much as Augustine's contemporaries would have recognized his self-characterization as Chaerea in

Eunuchus, so Harvey's colleagues would have recognized his self-characterization as Demea in *Adelphi*. This study explained how Harvey's quotation from that Roman comedy cued the plot of his physiological research and philosophical reasoning about the heart as the circulator of the blood. Harvey exploited the plot twister of all Roman comedy, error, to cast himself as a corrector. This study interpreted his citation of Demea's soliloquy to his colleagues because *Adelphi* (The brothers) involved family problems, a fraternal conflict about fathers raising sons, either strictly or leniently. Harvey appropriated that soliloquy to address the conflict in the College of Physicians about his own role as their anatomist now accused of a breach of oath to the medical tradition. The College paternalistically required under oath the fidelity of its members to the notions of ancient physicians, notably Galen, rather than to the practice of a contemporary anatomy. Demea's soliloquy in Terence's comedy embodied a father's radical change of behavior toward his son. By quoting it, Harvey declared his own radical change of mind from rigorous conformity to the medical tradition on the heart, indeed to the very concept of tradition. He invited his colleagues to imitate his conversion as their anatomist from being a hidebound disciplinarian to being a meandering explorer. As he developed his argument about his changing roles, however, Harvey moved from courtesy to aggression, declaring civil war within the College and marking his own decisive step toward the medical science of the heart. This confrontation revealed his comprehension and valuation of his inventiveness and his invention. The study demonstrated how Harvey's humanist rhetoric ventured beyond imitation to emulation in order to assert his own talent above and against other philosophers and physicians. Harvey argued for disrespecting authority in order to be innovatively truthful about the function of the heart. And he would redefine medical authority itself as not traditional but empirical.

"Aristotle's Cardiac Vessel" (new) reveals how that innovative natural philosopher used ordinary artifacts, the amphora and the oven, to postulate the movement of the heart as natural, meaning intrinsic. "Harvey in the Sluice: From Hydraulic Engineering to Human Physiology" demonstrated Harvey's methodical comparison of mechanics to nature as a heuristic for cardiovascular physiology. It reported how his discovery of the heart's circulation of the blood owed to the most valued invention after the wheel in the history of engineering, the hydraulic pound lock. Toward researching the movement of the heart and the blood, Harvey's anatomy experimented by cutting into a ligated arm and probing its venous membranes. He observed that those membranes all flattened forward under his probe but retracted upright upon its withdrawal. Although his professor Fabrici had supposed that they functioned to regulate proportionately the blood flow forward, like wickets in grain mills, Harvey posited that they functioned to block totally the blood flow backward, like sluice gates in water channels. Harvey reasoned toward a one-way blood flow circulating from and to the heart, like walking the unicursal design of a labyrinth to its center. This study located Harvey's observation of hydraulic engineering in London where he practiced anatomy, and in Padua where he had earlier studied it. It found that he personally experienced his cardiovascular model of the blood flow in voyaging through Leonardo da Vinci's attributed design of a double-mitered sluice gate, the Porte Contarine lock on the Brenta Canal. That Paduan prototype, together with the first pound lock in England under construction on the Thames River, provided him his technological model for the prevention of a fluid reflux of the blood in the body. Just as those double-mitered sluice gates prevented the backflow of water in a canal or river, the venous membranes obstructed the reflux of blood in the veins. This study explained that Harvey's discovery of the heart as the circulator of the blood was not entirely empirical since he did not actually probe instrumentally through the veins to the heart. His achievement was an interdisciplinary alliance of observation and reason, of tests and thoughts, in sum, of anatomy and philosophy.

"Harvey, by Hercules! The Hero of the Blood's Circulation" (reprint) reveals Harvey's ambition for immortality as the natural philosopher of the heart's circulation of the blood. This study unmasks his persona as Hercules, the superman of classical legends, by whom Harvey's book swore *mehercule*, despite current English law that forbade swearing. Its research exposes Harvey's rhetorical invention of the blood's circulation by his anatomical imitation of a particular Herculean labor, the cleansing of the Augean stable. Harvey's forbidden oath *mehercule* in his book marked his rejection of Galen's oath *Dia* in that physician's book, which had asserted a porous cardiac septum. Although Harvey was sworn as the anatomist of the College of Physicians, London, to teach Galen's medicine, he subversively assumed the persona of Hercules to embody his own anatomical labor. Harvey reprised that role in self-defense against accusations of his breach of oath with the medical tradition. He sought to usurp the medical epithet "a second Hercules" by reforming humanist dependence on ancient texts, such as Galen's, as authoritative medicine. Harvey's labor in reforming medicine compared with Hercules's labor of cleansing the Augean stable, a reformatory topic of Renaissance humanism. This study reports Harvey's decisive rejection of Galen's porous cardiac septum, which supposedly allowed a blood flow across the cardiac ventricles. Harvey's *mehercule* swore against Galen's *Dia* to assert the necessity of opening an alternate route for the blood flow. Herculean legends about cleansing the Augean stable interpret Harvey's labor to dam the cardiac septum and divert the blood flow into a continuous channel through the arteries and veins. His theory of that circulation further imitated Hercules's successful dependence on water flow to flush the Augean stable by applying liquid force to clean the medical stable. Harvey's term *copia* was functional, not quantitative. It did not denote a measured amount of blood but a powerful supply of blood. This final cultural anatomy of the heart exposes Harvey's professional issues and personal ambitions, toward a fuller understanding of his extraordinary role in historical understanding of the physical heart as the seat of the human soul.

Harvey's anatomical observations and reasoned arguments, composed with literary genius, have stabilized the far bookend for this new volume on the heart. The College of Physicians, London, acknowledged his extraordinary labors as the anatomist and philosopher of the heart by its exultant, if posthumous, bestowal on him of "immortality." Like the legendary Hercules, the man Harvey became a god. For, before Harvey's discovery, the movement of the physical heart and blood flow had seemed "almost known to God alone." In the beginning, in the Hebrew Bible, stood God's judgmental knowledge of human "heart" as the agent for or against his law. The association of biblical "heart" with the law devolved to his creation of humans "in our image and likeness" (Genesis 1:26). That phrase was unique in the Bible, but its original meaning was lost until a chance discovery in 1979 by a Syrian farmer bulldozing his field. He uncovered a life-sized basalt statue of a male whose skirt was engraved with the oldest extant Aramaic-Akkadian inscription. Its comparable terms "image" and "likeness" denoted the king's delegation of authority to the provincial official whom the statue represented. Biblical "heart" was thus originally associated with a legal relationship sealed in the Israelite covenant. Classical philosophers and physicians later associated or even identified the soul in blooded animals, most excellently humans, with the physical heart. Under their influence Christian theologians identified the biblical "image and likeness" as the human soul with its mental faculties of rational and desiderative movements. The movement of the physical heart was acutely examined in the classical era by Aristotle as a natural philosopher and in the early modern era by Harvey, his latter-day disciple but dissenter. Movement was a common denominator of both conceptual and physical hearts toward a philosophical final cause or a biblical divine Creator. Heart in those ancient to early modern cultures was the human agency for movement toward the divine source, whether by obeying God's biblical commandments, beating in harmony with his universal creation, reasoning in conformity with his imprinted natural law, discovering and dividing phenomena toward his final causality, receiving his spiritual renewal, or imitating a legendary immortal god. Those historical hearts evidenced essential aspects of human existence that still endure in modern thought and experience of political community, psychological mentality, and physical vitality.

Toronto, ON — Marjorie O'Rourke Boyle

Author's Publications on the Heart

"Aquinas's Natural Heart," *Early Science and Medicine* 18 (2013): 266–90.

"Augustine's Heartbeat: From Time to Eternity," *Viator: Medieval and Renaissance Studies* 38 (2007): 19–43.

"Broken Hearts: The Violation of Biblical Law," *Journal of the American Academy of Religion* 73 (2005): 731–57.

"Harvey, by Hercules! The Hero of the Blood's Circulation," *Medical History* 57 (2013): 6–27.

"Harvey in the Sluice: From Hydraulic Engineering to Human Physiology," *History and Technology* 24 (2008): 1–23.

"'In the Heart of the Sea': Fathoming the Exodus," *Journal of Near Eastern Studies* 63 (2004): 17–27.

"Pure of Heart: From Ancient Rites to Renaissance Plato," *Journal of the History of Ideas* 63 (2001): 41–62.

"Reprising Terence's Plot: William Harvey's Soliloquy to the College of Physicians," *Medical History* 52 (2008): 365–86.

"The Law of the Heart: The Death of a Fool (1 Samuel 25)," *Journal of Biblical Literature* 120 (2001): 401–27.

"The Wonder of the Heart: Albert the Great on the Origin of Philosophy," *Viator: Journal of Medieval and Renaissance Studies* 45 (2014): 149–72.

"William Harvey's Anatomy Book and Literary Culture," *Medical History* 52 (2008): 73–90.

Contents

Aristotle's Cardiac Vessel

"Come in; don't be afraid; there are gods even here."
Aristotle, *De partibus animalium*, citing a fragment from Heraclitus at the oven to visitors[1]

The heart, Aristotle stated, was a "vessel."[2] Although, from his famous dissections of the chick embryo, he observed the heart in motion "as if it were an animal" and "by nature a sort of animal,"[3] he identified the heart with an inanimate artifact. His term "vessel" has been subject to serious criticism. Aristotle's "fundamental model, the heart-as-a-container, was not physiological." And, "Aristotle's main model for the heart is simple and entirely structural: the heart is a container."[4] This study examines his cardiac vessel not systematically in the context of modern science[5] but historically in the context of Aristotle's culture. Greek art, of which pottery was an excellent type, was functional,[6] and the function of a pottery vessel was determined by its structure.[7] The actual structure and function of a vessel were inseparable. The usage of a vessel afforded Aristotle a heuristic and pedagogical model for the supply—not the storage—of the blood. For Aristotle, as for physicians, that blood supply was a matter of life and death.

M. O'R. Boyle, *Cultural Anatomies of the Heart in Aristotle, Augustine, Aquinas, Calvin, and Harvey*,
https://doi.org/10.1007/978-3-319-93653-6_1

Nature and Crafts

A Greek vessel was a common object but a civic pride. The contemporary philosopher Critias lauded Athens for the invention of "the potter's wheel, and the child of earth and oven, most glorious pottery, useful household ware."[8] The praise was a fragment and its claim was untrue,[9] although Attic pottery was indeed useful and sometimes glorious. Yet, from Greek antiquity no potter's wheel has survived intact, and even its depictions on the vessels it formed are few.[10] From the study of excavations of badly damaged pieces and again from scarce depictions, the Greek kiln has only recently been recreated and fired successfully in a modern backyard in Tampa, Florida.[11] And, although pottery survived better than any other ancient artifact, and the types of Greek vessels in modern classifications are numerous, an estimated less than one percent of their ancient production is extant, mostly as shards.[12] Fewer than half of Aristotle's own works on their earliest list are extant.[13] Those survived the vagaries of human transport and deposit[14] and the ravages of the very nature that incited his research. For the history of the heart, especially valuable were the lost anatomical drawings.[15] Those artisanal renderings would have intended to copy nature's design in the generation of animals. For Aristotle believed that "in the early stages the parts are all traced out in outline; later on they get their various colours and softnesses and hardnesses, for all the world as if a painter were at work on them, the painter being Nature." He cited the common observation that painters "first of all sketch in the figure of the animal in outline, and after that go on to apply the colours."[16] His personal preference was for black-and-white line drawings for design rather than colored pictures for character.[17] The starting point for Nature's design of blooded animals, the noblest of its productions, was precisely a point, the drop of blood Aristotle observed in a chick's embryo at three days.[18] That evidence survives in a text but not an illustration. The remnants of Aristotle's ancient culture invite research behind the scenes to find and to understand better his cardiac vessel. For, an alternative to the regret that Aristotle was ignorant or inexperienced about cardiac physiology is the premise of this study that he did not explain certain facts because their usage was so ordinary and obvious to his contemporaries that they needed no explanation. A vessel was such a fact. As an introduction to the subject states: "In ancient Greece vases were commonplace. No one needed to be told what they were used for or what their pictures meant; pottery was just part of everyday life. Ceramic vases served as containers, either for utilitarian purposes in the home or for religious rituals. Some were decorated with figures and patterns, others were plain and painted black all over; unpainted coarse ware sufficed for cooking."[19]

Aristotle's master in philosophy, Plato, categorized a "vessel" or "container" (*angeion*) as the prime product of the human skills that contributed to the civic good. A vessel was the class of things made with tools "for the sake of preserving what craftsmen have produced." Plato identified "this varied kind of thing which is worked for things liquid and solid, and for things that are prepared on the fire and things that are not, and which we refer to with the single name of 'vessel'—a common kind of thing."[20] A vessel provided Aristotle an analogy for the heart from what man made by art to what was made in man by nature. His deliberations referred from nature to industry so as to compare animal heart to human artifact, the vessel. The comparison was not an equivalence. Animals and their organs existed by nature. As Aristotle defined their nature, they "have within themselves a principle of movement (or change) and rest—in some cases local only, in others quantitative, as in growth or shrinkage, and in others again qualitative, in the way of modification." In contrast an artifact possessed no such internal inclination toward change. Yet, by its composition of stone or earth, or a mixture of those, it incidentally possessed the principles of change inherent primarily in its materials as natural substances. Aristotle's example of incidental possession was the physician healing himself. "If a man were a physician and prescribed successfully for himself, the patient would cure himself; but it would not be *qua* patient that he possessed the healing art, though in this particular case it happened that the physician's personality coincided with that of the patient, which is not always the case." Aristotle reasoned, "And so it is with all manufactured or 'made' things: none of them has within itself the principle of its own making." That principle resided in "some external agent,"[21] such as a potter.

Aristotle's artisanal vessel was an ordinary possession. It was displayed in the stalls of the Agora, or central marketplace,[22] and accessible in the nearby potters' quarter.[23] Athens was renowned for the discovery and development of the red-figure technique of vase painting, which became the norm.[24] However, personal artistry in a more colorful but less defined style peaked by the middle of Aristotle's century, and by the year of his death in 322 B.C.E. painted vases were no longer being made.[25] Aristotle left no comment on the vase paintings of the mythological and mundane scenes that scholars and connoisseurs now admire, although he cited the proverbial rivalry of "potter against potter."[26] Painted vessels were not essential to Aristotle's artisanal model for the natural heart; undecorated vessels sufficed. Coarse ware was vastly more common than fine ware[27]; on most painted vessels the servers and drinkers handled undecorated ones.[28]

Aristotle's model of the vessel coordinated with animal research. He reported a vessel lowered into the sea by a cord as an experiment for how marine animals survived in salty water with access to sweet water for concocting bodily parts and feeding offspring.[29] His records of vessels for generation offered, perhaps, a rare glimpse into his own practice. Aristotle indicated that warmed pottery vessels could serve as incubators for bird eggs.[30] Since his knowledge of the heart originated in dissections of the chick embryo, he may have used such vessels as incubators. He equated the spoiling of egg yolks incubated in hot sunny weather with the souring of wine, which was normally put in pottery vessels.[31] He also reported pottery vessels discarded at sea, in which the octopus deposited and concocted its eggs. An egg was so large that it could fill another vessel the size of the octopus's head.[32] Two vessels implied a transfer, with the second used for research. Aristotle did inspect the interior of the octopus's egg to describe its fluid.[33] He also recorded the chance use of a vessel for gestation. He told of a pregnant mouse accidentally shut up in a vessel of millet. When the vessel was opened, a mischief of 120 pups appeared.[34]

Aristotle knew the technology for vessels. The potters' quarter (Kerameikos) of Athens, the prime site of production, straddled the banks of the Eridanos River near the main Dipylon Gate on the northwest boundary.[35] Aristotle passed through the potters' quarter for twenty years in traveling the Dromos road to and from Plato's Academy, which lay 1.5 km north of the city.[36] Plato's philosophy compared the process of education toward responsible citizenship to the training of a potter from apprentice to master. He warned beginners not to undertake a large jar, a pithos, foolishly. As Socrates asked, "Did you never observe in the arts how the potters' boys look on and help, long before they touch the wheel?"[37] A pithos that could fit a man inside was not even turnable on the wheel, thus molded by hand.[38] Aristotle's inspection of Athens's pottery yards and workshops[39] would have afforded him empirical evidence of material causes and compounds. His early dialogue *Protrepticus* voiced his conviction. "It is necessary much earlier to be intelligent about the causes and the elements than about the posterior things … It is out of the former and because of the latter that the other things come into being and are evidently constituted." Concerning the elements or other natures, "it would be impossible to be mistaken about these things and recognize any of the other things."[40]

Those basics were the elements of earth, water, fire, and air with their qualities of hot, cold, dry, and wet.[41] The elements of earth and water formed the compound clay, while the elements of fire and air in a kiln transformed that raw material into durable pottery. The potters' technology was the efficient cause, "the source of movement or change,"[42] for the perfected vessel. Attic clay, among the finest in the world, was sedimentary, with deposits of iron that gave Greek vessels their characteristic red-orange color.[43] Aristotle identified potter's clay (*keramos*) as a compound of earth and water, predominately earth, that was insoluble.[44] His identification followed another compound of earth and water, also predominately earthy, the blood. Such compounds solidified and densified either by cold or by heat[45]: earth by heating, blood by cooling.[46] Fire acted on compounds of earth and water "like clay when baked." As he explained the process, "Heat draws out the moisture, and when the moisture evaporates the dry constituents increase in density and pack together." Soft and dry things "solidify, like clay when baked." Aristotle detailed the process in the kiln. "Bodies which have been made dense or hard by cold often become moist at first when heated, like clay again, which when baked steams at first and becomes softer (which is why it sometimes becomes distorted in the kiln)."[47]

As for unbaked clay, its porosity provided Aristotle a comparison between a pottery vessel and a blood vessel. He explained the formation of bodily flesh as "the nourishment oozes through the blood-vessels in the several parts (just as water does when it stands in unbaked earthenware."[48] The mature product was completely dried clay, his cardiac model, the source of the blood for distribution through those blood vessels. Aristotle observed, "Earthenware (*keramos*) is composed of earth only because when dried it solidifies gradually; neither can water gain entry through pores from which only vapour could escape, nor can fire, which was the solidifying agent."[49] However, Greek kilns heated only to about 950 degrees, whereas nonporosity required a temperature of 1200 degrees and above. The technique of a siliceous vitreous glaze to seal pottery watertight was unknown in classical antiquity.[50] Greek vases were only superficially coated with a clay slip.[51] For security in cartage or shipping, the insides of the coarse transport amphoras were frequently lined with 1–2 mm of pitch.[52] Despite some residual porosity, Greek earthenware, when correctly fired in the kiln, did not leak or seep liquid contents nor did it crack or break when reheated over the fire as a cooking vessel. Those properties—impermeability to liquids and resistance to fire—made the earthenware vessel an intelligible model for Aristotle's heart as a container of the blood supply and preserver of the vital heat. As he stated, "The middle of the heart is a body which is naturally dense and hollow; and further, it is full of blood, inasmuch as the blood vessels originate there; it is hollow to serve as the receptacle of the blood and dense in order to guard the origin of heat."[53]

Container and Contents

Aristotle's attention to the elements and qualities based his cardiac vessel of the blood. It was because the blood was moist that it needed a container, for "all natural bodily fluids belonged in vessels." The blood as a fluid was thus "always in a vessel, in those called blood vessels (*phleps*) and not otherwise except in the heart alone."[54] Aristotle's diction for "vessel" or "container" was Ionian *angeion*,[55] the standard dialect for medical, scientific, and historical texts.[56] In the Hippocratic Corpus before Aristotle,[57] *angeion* was negligible as a term. Although a therapy to alleviate a side ache was a hot-water bottle, whether that container was a wineskin, a pouch, or a pottery or copper vessel (*angeion*) was indifferent.[58] A cure for barrenness specified earthenware. A handful of horehound should be placed in an Attic vase (*angeion Attikon*), then four cups of water poured in to steep the potion.[59] A treatise on diseases used *angeion* for vessels from the spleen, the source of humoral water, although it otherwise used *phleps* for a vessel.[60]

Aristotle's bodily *angeion* occurred principally in *De partibus animalium*, an investigation of causes, and *Historia animalium*, a record of research and data.[61] As a natural philosopher, he analyzed the vessel philosophically. *Protrepticus* early stated the usefulness of philosophy for medicine, citing the agreement of "sophisticated doctors" that "good doctors … must be experienced about nature."[62] *De iuventute et senectute* concluded with the reciprocity of philosophers to physicians for medical principles. Their investigations overlapped some because of a shared interest in the causes of health and disease.[63] As Aristotle's interdisciplinary commitment has been well understood, it "posited a continuum between doctors and natural philosophers: most natural philosophers in their account of the world had also to consider medicine, while the more thoughtful physicians grounded their medical theories upon the principles of natural philosophy."[64]

Aristotle's logical *Categoriae* defined "vessel" (*angeion*) in a survey of the ordinary meanings of the verb "to have." The survey concluded several miscellaneous chapters, the *post-praedicamenta* after the categories. "*Having* is spoken of in a number of ways … as in a container as with the measure of wheat or the jar of wine (for the jar is said to have wine, and the measure wheat, so these are said to have as in a container)."[65] The definition of "to have" in *Metaphysica* agreed with that notion of the vessel holding a liquid. "We speak of anything as 'having' in which, as receptive material, something is present." For example, "The body 'has' the disease." For "the sense that the container holds the contained; for when A is contained in B, we say that A is held by B. E.g., we say that the vessel holds the liquid."[66] Those concise mentions tacitly allowed "having" to be subject to change by external agents, a steward or a physician, for wine to be poured from the jar into a drinking cup and for disease to be cured.

Aristotle's *Physica*, about nature, considered the vessel (*angeion*) intently and significantly. It deliberated about containers and contents in its discussion of place as essential to motional change.[67] Replacement, the local change by which something exited from a place such as liquid from a vessel, then something else such as air entered the vessel, established the difference between container and contents.[68] The successive contents of liquid then air were separable from their place in the vessel. "In fact a 'place' seems to resemble a vessel, a 'vessel' being 'a place that can itself be moved about.'" Aristotle reasoned, "Just as the vessel is no part of its content, so the place is no part of that which is in it."[69] He considered the different meanings of something being in something else. One use was the inclusion of the part in the whole, such as the finger in the hand, or conversely the whole as comprising all its parts. Another use was the inclusion of a species in a genus, such as "human" in "animal," or conversely a genus in a species. Another use was that "health may be said to 'have its seat in' the balance of warm and cold." And so forth. "But the primary sense, from which all these are derived, is that in which we say that a thing is 'in' a vessel, or more generally 'in a place.'"[70]

Aristotle asked whether "a thing can be in itself or whether it must always be in something other than itself, if it be anywhere at all."[71] Was a content in a container primarily and directly, or relatively and indirectly? His case was an amphora (*amphoreus*) of wine. He decided that neither the amphora nor the wine was in itself, but that the amphora of wine could be in itself because both were parts of the same whole. He adduced from health the parallel of the pallor of a person's skin extended to the entire person as pallid, like wine in the amphora not as separate entities but as a whole. Aristotle argued that nothing could be in itself by definition because each of the separate parts would have to be defined as both; namely, the amphora as both amphora and wine, and the wine as both wine and amphora. However, the amphora received the wine not as itself wine but as a container, and the wine was in the amphora not as itself amphora but as contents. The difference between container and contents was obvious to him. Nor could a thing be in itself coincidentally because the amphora and the wine would have to be the same, which was impossible. Aristotle concluded, "The impossibility of anything being, in the primary sense, in itself, is clearly demonstrable." He determined that "the vessel is no part of its own content."[72]

The amphora of wine was his model for the heart as a vessel of blood. *Physica* established the distinction between container and contents by analyzing the local movement of replacement, by which the content of a vessel was emptied from it to another place. That reasoning about place was fundamental to the identity of the cardiac vessel as distinct from its content of blood. The distinction allowed importantly for the displacement of the blood from the heart. Aristotle's philosophical deliberation on place for a change of content clarified that his physical research on the heart was about a change of place for its contents. The cardiac vessel was not a static but a dynamic model. A vessel, in the ordinary Greek usage that based his reasoning, did not store its same contents forever. It held them only until the vessel needed to be emptied for their consumption. Actual household usage was fundamental to Aristotle's meaning of the vessel. He partly defined a house as a protective vessel sheltering goods such as vessels.[73] Since possessions belonged to a household, he thought its management required skill in acquiring and using those necessities for living and living well.[74] Storage of food was a considerable project since many Greek households intended to have a year's supply. That quantity required spacious and specialized facilities to preserve dry goods from vermin and liquids from air. A household of four to six persons has been calculated to need hundreds of liters of wine annually, more if it made its own wine of a kind to be aged.[75] Aristotle philosophized particularly about place to refute the notion of the void, by the example of the pithos,[76] the larger vessel for fermenting the new wine. A pithos ranged from about 0.9–1.0 m high with average capacity of 100–350 liters to more than 2m high with a capacity of 1000 liters or more.[77] It named the opening day, Pithoigia, of the Athenian festival Anthesteria to open, decant, and consume the new wine.[78] Aristotle philosophized generally about place with the example of the amphora of wine. An amphora held from 7 to 75 liters,[79] with most amphoras ranging from 15 to 25 liters.[80] A domestic amphora, the usual wine decanter, was about 30–45 cm high.[81] It was comparatively liftable for pouring its contents into the krater, the wide-mouthed vessel where the wine was mixed with water, then ladled out for serving.[82] An amphora was a standard Greek measure for liquid volume.[83] For example, Aristotle reckoned the yield of goats' milk versus cows' milk for cheese making in amphoras.[84] Both a pithos and an amphora only held their contents temporarily until needed for consumption. Wine could sour; oil and milk could go rancid. Grains could become infested with pests—or a pregnant mouse. Aristotle wrote that while an amphora of wine was not as real as a friend, he wished it safe keeping until he was ready to enjoy its contents. As he said, everyone enjoyed wine with their meals.[85]

Like the vessels of wine, the heart as a vessel contained the essential nourishment of blood only temporarily until distributed for consumption. The heart had blood but not always the same blood. As Aristotle observed in the chick's embryo, the heart was immediately blooded,[86] before the differentiation of the blood vessels.[87] As the origin of those blood vessels, the heart "has within it the primary potential for fashioning blood (*echein en autē tēn dunamin tēn dēmiourgousan to haima prōtēn*)."[88] The heart was like a productive workshop (*dēmiourgeion*). It had "a continuous influx of this fluid, of which the blood is constituted; for it is in the heart that blood is first manufactured."[89] Because the heart was a sufficient producer it had no need to store blood. It continuously made blood and continuously emptied it into the blood vessels for distribution. The heart was not a storage but a supplier, not inert and passive but initiating and active. The heart was an agent of change, and the primary change was movement, that is, change of place for the blood.[90] The heart moved its contents of blood from itself as a container into the auxiliary blood vessels. Aristotle termed the heart a "vessel" basically because all bodily liquids needed containment[91] lest they flow indiscriminately. The heart contained blood not to hold it, however, but to hold it *in*. Of the uniform parts of blooded animals, some such as blood were moist, while others such as the blood vessels and the heart were solid.[92] The heart delimited the liquid blood by the solid boundaries of its exterior walls and interior chambers. That container was a controller, to direct the movement of the blood purposefully and effectively. The heart had blood in order to channel it into the blood vessels for distribution. "Blood is present in blooded animals for the sake of nourishment, i.e. nourishment of the parts."[93]

Aristotle's first task for articulating the heart as a vessel was to eliminate any confusion of container and contents. He criticized the notion that generation happened from a vessel, rather than from some matter.[94] Aristotle thought that in the embryo all parts of blooded animals were formed from bloody matter. The heart, as the origin of the blood, was "the primary blooded part." It was as such "immediately blooded."[95] As the origin and source of the blood, the heart was constituted by blood.[96] It was necessary, therefore, to distinguish that vessel made from blood from the blood it contained. Aristotle also distinguished blood from the flesh that it formed as a bodily constituent. "In fact blood is not continuous with flesh, nor united with it but lies in the heart and in the blood vessels as in a container."[97] Aristotle's next task was to distinguish the heart as the primary vessel for the blood from the auxiliary blood vessels. For, he called both *angeion*[98] although he usually named the blood vessels *phleps*, as traditional. The blood vessels existed for the blood, "for that which is entirely moist has need of a container (*angeion*), and the kind consisting of blood vessels is a container (*angeion*), and the blood is in these."[99] Aristotle preferred one source for the blood, in the central heart, rather than in its networked blood vessels, as the unit for access to and control of the blood. His empirical argument, from dissections, for the heart as the origin and source of the blood was that no blood vessels transversed it. "And the heart is the origin of the blood vessels; for they are clearly from the heart and not through it, and its nature is vascular as if it were like in kind to them." Nature contrived the blood vessels relative to the heart, by complement and juxtaposition, lest the heart suffer from its own excess of blood and of heat.[100] Aristotle thought that the first thing that held the blood, as in a vessel, necessarily was its source. Since the blood went from the heart to the blood vessels, but not vice versa, he designated the heart "an origin and spring of the blood, or its first receptacle."[101] His phrase "origin and spring" otherwise appeared only in his analysis of the sources of political sedition, the causes of the breakdown of the unified civic body into factions.[102] The heart was the central unit whose basic division into the great blood vessel and the aorta allowed the blood to flow into the many further divisions of the blood vessels.

Anatomy and Vessels

Aristotle's *Physica* distinguished container from contents by the local movement of replacement. His model for that change, the amphora of wine, served for the cardiac vessel of blood. Greek vessels were described anatomically, so his analogy of craft and nature was very familiar. Classical taste preferred vessels with "tense, muscular shapes." Modern nomenclature has copied bodily parts: mouth, lip, neck, shoulders, belly, and feet.[103] The Greeks wrote of the head of the vase, its mouth, its lips, its belly, with sometimes a navel; of the inside of a cup as a face; and of dual handles on vases and cups as ears. Vases were molded after a human face, or the female breast, or the male genitals. Vases were also molded after animal heads, wild and domestic, or after a horn, claw, or hoof.[104] In sum, "In the hands of the potter, the vase resembles a body to which he gives shape."[105] Aristotle wrote that "the material for generation must be immediately available in the female just as the potter is close by his clay."[106]

Aristotle named three bodily parts a "vessel" *(angeion)* for liquids: bladder, breasts, and heart.[107] All of these vessels emptied their fluids. The bladder was a vessel that collected the excessive residue in lunged animals. The kidneys assisted by passing that residue into the bladder through attached channels.[108] The flow between those organs moved downward toward excretion, as consistent with bodily structure. "The part below is the place for the nourishment and the residue."[109] The female breasts were vessels that supplied nourishment for offspring.[110] Their milk was the surplus residue from the formation of the fetus or embryo.[111] As a compound of earth and water, milk tended down.[112] Aristotle detailed the places of the breasts in other animals down where nourishment and residue collected.[113] But he did not explain how in humans the residue for concoction as milk exited the uterus below the diaphragm and flowed up to the breasts above the diaphragm. He stated, "It makes its way out, and changes over to another process of formation ... The milk collects in the upper part of the body, in the breasts, and this is accounted for by the original order of the body's construction." As he identified that, "The part of the body above the diaphragm is the controlling part of the body."[114] The upward flow of human milk from the uterus to the breasts deferred to their function in both sexes. The breasts protected the area around the heart,[115] which controlled the body from above.[116] By similitude Aristotle associated the breasts as a "vessel" (*angeion*) with a "receptacle" (*hypodochē*).[117] The proper bodily receptacles were the esophagus, the stomach or crop, and the intestines.[118] Aristotle stated that the heart had in its center at least one hollow cavity (*koilia*) as the initial receptacle (*hypodochē*) of the blood.[119] "For there must be a certain place within the heart, i.e. a receptacle (*hypodochē*) of the first blood."[120] That place was its hollow middle,[121] which in larger animals could be sectioned into two or three cavities.[122] The cavity (*koilia*) was a vessel (*angeion*)[123] within the vessel.

Aristotle began on the heart precisely at a point. That argument conformed not only to empirical observation but also to correct definition through terms that were prior, for example, a point before a line.[124] Both the heart and the liver appeared in blooded animals with their constitution. "In fact they are sometimes apparent in three-day old eggs, the size of a point, and very small ones are also apparent in the aborted remains of embryos."[125] Although both organs were a point (*stigmē*), Aristotle eliminated the liver as the starting point (*archē*) of the body. It lacked a receptacle (*hypodochē*) for the blood, only having its blood in a blood vessel.[126] Aristotle observed in the embryo that "a bird's development began from the pointed end."[127] As he amplified, "During this time the yolk has worked its way up towards the pointed end (*oxu*), where the 'principle' (*archē*) [starting point] of the egg is situated and where the egg hatches; and the heart is no bigger than just a small blood-spot (*stigmē*) in the white."[128] The "blood-spot" was at the "point," or farthest end (*akrōn*) of the heart, which was "sharp (*oxu*)."[129] The developed heart was shaped with a rounded top and its pointed end forward.[130] Aristotle observed the embryonic "point" (*stigmē*), then he termed it *sēmeion*, a mathematical "point" or a physical "boundary, limit." He distinguished those disciplinary perspectives. Mathematics studied the point in abstract from the physical conditions of bodies. Natural philosophy, which was Aristotle's interest, studied the point with respect to physical entities in movement.[131]

The cardiac point coordinated with a specific type of amphora, which was Aristotle's model in *Physica* for deliberating on movement as change of place. The amphora was very familiar as dining servers, athletic prizes, and grave markers; in modern archaeology the amphora is the commonest find in sites and shipwrecks.[132] Among the types of Greek vases, cups, and jugs for holding and pouring liquids, the amphora was classified by its particular shape. It had a narrow neck, two vertical handles (*amphi* "on both sides" and *pherō* "to carry"), broad shoulders, and a bulbous body that tapered to a base.[133] The amphora whose shape resembled the heart was the commonest type, the pointed amphora. It was undecorated coarse ware that tapered distinctively to a point,[134] without the delicate standing feet unique to Attic pottery.[135] During manufacture the support for throwing the body on the potter's wheel was cut to a point or knob in turning.[136] That knob or "toe" was handy for grabbing the vessel from the bottom to lift it, although some examples were simply pointed. The design enabled efficient handling and shipping in bulk. In ships pointed amphoras were stowed in alternating rows, with their points in the upper row inserted between the necks of those amphoras in the row below in a zigzag pattern.[137] In households the pointed amphora was stabilized in a ring stand, or stashed in a corner, or set in a hole in the ground.[138] The smaller size could be balanced for carrying on a shoulder or a pole.[139] A pointed amphora was the popular choice both for transport,[140] from Athens as the prime site for the export trade,[141] and for local provision.

The amphora of wine in *Physica* provided Aristotle a heuristic model for how the cardiac vessel held its content of blood as essentially moveable. But how did the cardiac vessel discharge its blood into the tributary blood vessels for distribution? Aristotle identified the cardiac outlets as the great blood vessel and the aorta, which "receive" the blood from the heart.[142] How did the blood arrive at those outlets from the cardiac cavities below them? Ascent of the blood from the heart into its aorta was complicated by Aristotle's inconsistent locations for it. The aorta was either attached to the middle cavity, of medium size with the purest blood[143]; or it was attached to the left cavity, the smallest and with the coolest blood.[144] Ascent of the blood from the heart into the great blood vessel was consistently from the largest, right cavity with the hottest blood.[145] Aristotle named the directions of bodily places and movements: above and below, front and back, right and left, and he designated the honorable places as above, front, and right.[146] He argued for the primacy of the heart from its place. "It is situated in an originative place; that is, it is near the middle, and more above than below, and more in front than in the rear; for nature places the more valuable things in the more valuable locations, where nothing greater prevents it."[147] Aristotle asserted not only the reality but also the influence of place. "The trends of the physical elements (fire, earth, and the rest) show not only that locality or position is a reality but also that it exerts an active influence." Fire ascended, earth descended, with each element tending to its proper position, as a division or class of general location. The directions up and down, right and left, in front and behind were not, as applied to the elements, relative to the position of the observer. "In Nature each of these directions is distinct and stable independently of us." For the vertical dimensions, up was the tendency of the lighter elements—air and fire, down the tendency of the weightier ones—water and earth.[148]

Aristotle thought blood was a compound of earth and water,[149] which elements did not naturally tend up—from the cardiac cavities into the great blood vessel and the aorta above it. How, then, did the blood ascend in and exit from the heart? Aristotle's heuristic heart was materially coherent because a pottery vessel, a fleshy vessel, and their respective contents of wine or blood were all compounds of earth and water, which tended down. An amphora of wine was emptied by an external agent. Artistic scenes of symposiums depicted a servant or a satyr gripping the handle of the amphora and hefting and tilting it to pour its wine into the krater.[150] The larger, heavier transport amphoras featured the knob for grabbing the vessel on its bottom to invert it.[151] The use of a siphon has also been proposed.[152] A *siphōn*, "tube, pipe," was "used for drawing wine out of the cask or jar."[153] The verb *siphōnizō* meant "to draw off wine with a siphon." It occurred uniquely, however, in Aristophanes's comedic *Women at the Thesmophoria*, to ridicule Athenian women who slyly siphoned wine from their household vessels. By the time of Aristotle's most respected commentator, Alexander of Aphrodesias in the late second or early third century C.E., the practice would be clear: "by sucking air out of the siphon we drain off the wine."[154] But Aristophanes's verb mocked silly female sipping through straws versus sociable male drinking from cups, the Athenian custom. Aristotle was the first thinker to mention mechanics as a discipline, and the first *Mechanica* was among his collected works, although inauthentic. It did not include the siphon.[155] Aristotle did not entertain the great blood vessel and the aorta as siphons to suck blood from the heart as their "origin and spring."[156] He rejected any instrumental regulation for a spring of water to flow from its source like a wine steward dipping a jug into a krater. "We must not understand a source from which waters are ladled as it were from a vessel, but a first point at which the water which is continually forming and percolating gathers."[157] The heart was the "spring" of blood, from which the auxiliary blood vessels carried it, like the watercourses in gardens.[158] There was a public model. On the Peripatos level of the Acropolis was the Klepsydra house for the spring that flowed down into the gardens, in the modern excavation record coincidentally called "the very heart of primitive Athens."[159]

Aristotle's agenda for causal investigation of the heart was "what sort of thing it is, what it is for the sake of, and the cause owing to which it is present in those animals that have it."[160] That account did not treat how the blood moved from the heart into the blood vessels; it only indicated that it did so. "The blood is conducted from the heart and into the blood vessels."[161] Aristotle acknowledged that the how of an action was harder to know than its fact or necessity, and he did so in a comparison between ethics and medicine about acting justly and prescribing effectively. "But how the just things are done and how they are distributed—this is indeed a greater task than to know what is conducive to health, since even here to know about honey, wine, hellebore, cauterizing, and cutting is easy." That comparison prioritized a requirement for all humans over one for some humans, the physicians. However, Aristotle continued by acknowledging medical skill equal to medical knowledge. "But to know how one must administer them with a view to health, and to whom and when, is as great a task as to be a physician."[162] As he asserted practically, "We are not healthy by being acquainted with what produces health, but rather by applying it to our bodies."[163] Aristotle's practical thoughts on the how blood emptied from its cardiac vessel were piecemeal, like shards. The cavities of the heart, which supplied the blood, were positioned lower than the great blood vessel and the aorta into which they delivered the blood up. Blood was a compound of earth and water, which elements tended down. The contrary elements were air and fire, which tended up. Of the causal factors in the elements, hot and cold were active, moist and dry were passive. "It is always heat and cold that are observed to determine, combine and change things both of the same and different kinds, as well as moistening, drying, hardening and softening."[164] He stated, "All heat naturally rises." Again, "Heat when radiated disperses into the upper region" for it has a "natural movement upwards."[165] In blooded animals the source of their vital heat was the heart.[166] It was specifically the origin of the heat and moisture in the blood.[167] Heat was necessary to maintain the blood as moist for its very flow because cold solidified blood, unless it was too watery.[168] The cardiac vessel functioned not only to supply blood to the auxiliary blood vessels but also to protect the vital heat without which the blood could not flow there. "The middle of the heart is a body which is naturally dense and hollow; and further, it is full of blood, inasmuch as the blood vessels originate there; it is hollow to serve as the receptacle of the blood and dense in order to guard the origin of heat."[169] The amphora served as a model of the cardiac vessel for the supply of the blood. As a model for the movement of that blood from the heart into the other blood vessels, Aristotle appropriated another vessel. This auxiliary vessel related to the other function of the heart, not as the source of blood but as the source of heat.

Hearth and Ovens

The heart was necessary "because there must be an origin of heat (for there is need of something like a hearth, in which lies the spark of the animal's nature, and that it be well guarded, being as it were an acropolis of the body)."[170] Aristotle modeled the centrally located heart, as the guard of the body's vital spark, on the centrally located hearth of architecture. In Greek mythology Zeus assigned the goddess Hestia ("hearth") to the middle of the house, where to keep its fire burning she received the fatty portions of animal sacrifice.[171] Fixed hearths were a feature of some classical Greek houses, especially in the north including Aristotle's home town of Stagira, but also in Athens on the Acropolis slope.[172] The hearth culturally symbolized the basic social unit, whose rituals around its fire bonded the kinship group.[173] Aristotle compared civic and bodily organizations[174] and noted the office for sacrifice at the civic hearth.[175] He compared the heart heating the body with the hearth heating a room. The cardiac cavities and the blood vessels in the largest animals were colder relative to those in the smaller animals. That relativity was like the variable effect of a fire of the same size in a large or a small room.[176] Aristotle's *De anima*, on the soul, reported Heraclitus's belief in a "warm exhalation" as the first principle of the universe.[177] Heraclitus's anecdotal warming himself at an oven was not about gods in the kitchen,[178] however, for there was no such defined and fixed place. Greek ovens and cookers were characteristically mobile.[179] Heraclitus said, rather, that the gods were with fire anywhere. *De anima* had also reported Anaxagoras's belief that "the cause of beauty and order" was found "in all animals, great and small, high and low."[180] By conflating their ideas, Aristotle interpreted Heraclitus's invitation at the oven to mean to study all animals so as to learn about the vital heat, with its variable effects. Aristotle introduced his comparison about heating rooms of different sizes with the different sizes of hearts as large, small, or medium. Conditions in household rooms and cardiac vessels he declared "similar." He added that size also influenced temperament, such as fear in those animals whose "heat in their heart is not in balance (being small in quantity, it is weakened in large animals), and because the blood is colder." Large-hearted animals were "timid or devious," the typically scared rabbit or mouse.[181]

Aristotle's inclusive study of animals transcended the aesthetic norm. *Poetica* stated about the composition of tragedy, "Any beautiful object (*to kalon*), whether a living organism or any other thing made up of parts, must have those parts not only in proper order but also on an appropriate scale. Beauty (*to kalon*) consists in scale as well as order." He reasoned, "That is why there could not be a beautiful organism (*kalon*) that was either minuscule or gigantic." Either a quick glimpse of a tiny animal was indistinct, or a single glance at a huge one could not comprehend its whole. Such extremes of animal size disallowed the recognition and appreciation of proportionate parts in the whole.[182] Aristotle's causal study of animal parts radically altered that aesthetic judgment about the necessity of proportionate magnitude for natural beauty. As his citation of Heraclitus at the oven concluded, "So too one should approach research about each of the animals without disgust, since in every one there is something natural and good (*tinos physikon kai kalou*)."[183] By comparison of his texts, *kalos* there meant "beautiful." Aristotle decided that the study of animals should include all types without aesthetic discrimination. For, in nature purpose predominated, and the purpose of the things nature made belonged to "the beautiful."[184] Without prejudice he even scrutinized the bookworms that might destroy his scientific papers, "tiny animals" engendered "in books, some of them similar to those found in clothes, others like tailless scorpions, very small indeed."[185] In Aristotle's natural philosophy, purpose, overriding scale, informed his account of the heart being visible first as a tiny point.[186]

Cardiac heat, as if from a hearth, was necessary for life in blooded animals.[187] It effected concoction, the intrinsic process of maturation. Concoction was a major, if variable, term in Aristotle's animal books, especially for generation.[188] He studied the domestic culinary methods of ripening, boiling, and roasting seriously[189] for their transformations by heat. For the blood flow, he also studied the behavior of liquids. They were all designated as compounds of earth and water but classified by the predominant element. Wine was Aristotle's difficult case because its behavior varied with different methods of processing grapes.[190] However, he thought wine was usually more watery, unless concentrated by boiling the must.[191] Blood was usually more earthy, unless serous from disease or in animals lacking its fibers.[192] The heart supplied the nutritive blood and protected the vital heat for its flow. The amphora of wine served Aristotle as a model for supply. But he needed an auxiliary model, cookware, that was especially refractory to heat. Greek cookware was made from clay mixed with powdered fired clay or with volcanic rock. It was usually fashioned manually by coiling thick strands around a base and then smoothing the joins. The material and the method produced a rough but sturdy vessel that was even more fire resistant than other kinds such as the amphora.[193] The ordinary oven was the *ipnos* for baking. As attested by terracotta figurines of bakers and by actual specimens, it was a portable clay vessel set on legs, with a fire underneath that was stoked and tended. The oven was open in the front for placing the dough, and it had a slit at the rear for a draught.[194] The *ipnos* occurred in Aristotle's extant works only in his unique citation of Heraclitus warming himself at that oven.[195]

Aristotle's own model for the heart as the origin and preserver of the vital heat was a simpler oven, the *pnigeus*. It was a portable earthenware dome with a handle on top. Lighted coals were set on a cleared dirt floor, the dome was placed or hung over them; and, when the dome was sufficiently hot inside, it was lifted, and the coals were removed. The dough was then set on the warmed dirt and the cover fitted over it. Coals were heaped about the dome, and the bread baked.[196] Like the amphora, the *pnigeus* was formed from earth and water then fired in the kiln. When the *pnigeus* was set on the ground, the earth from which it was made became its temporary bottom. The *pnigeus* had Athenian notoriety in Aristophanes's comedies. *Clouds* mocked sophists as "quacks," like Socrates "who in speaking of the heavens persuade people that it is an oven, and that it encompasses us, and that we are the embers."[197] For animals, Aristotle compared a *pnigeus* to the domed shell of crabs, crayfish, oysters, and tortoises, which enclosed their flesh. As bloodless animals, their natural heat was low, so their shells acted "like a *pnigeus*, a surrounding pottery vessel (*perikeimenov ostrakov*) to guard the heat for baking inside."[198] He observed that the tortoise's shell allowed it to move its feet even after death because it conserved some residual heat.[199]

For Aristotle, animal life required heat. Whether death was violent by external cause or natural by internal cause, its universal cause was "a failure of heat." In blooded animals that failure happened in the source of its substance, the heart, which preserved the vital heat until it was either extinguished by an excess of its own heat or exhausted by the cold. He thought that, because most of the bodily heat was used up during a lifetime, a strain on the heart in old age quickly caused its extinction, like snuffing a low flame. Because the heat was no longer mitigated by the lungs, it burned itself out.[200] For continuity of life, there had to be a means of preserving the vital heat. Aristotle illustrated his reasoning by the operation of the *pnigeus*, the portable domed oven. His declared "paradigm" to get a handle, figuratively, on the cause of death was cooking in that oven. He stated that if coals were heated in the *pnigeus* unremittingly they were quickly extinguished. But if that oven was frequently and alternately lifted and settled, the coals stayed cooking for a long time.[201] A *pnigeus* had no vent[202] for regulating heat and air. It was operated by grasping its handle on top to move it. Aristotle's account of rapid alternations was based on his belief that fire could consume itself by excess.[203] So as not to exhaust itself, fire needed a preventative coolant, air. He believed that all blooded animals required rapid refrigeration of their cardiac heat lest it flare out of control.[204] A natural agent needed to come to the rescue, like the potter in a painting who rushed into the scene to adjust the wildly flaring chimney of his kiln.[205] Such cooling of the heart was a function of the lungs.[206] Inhalation and exhalation of air rapidly raised and lowered the chest,[207] just as Aristotle's *pnigeus* was rapidly manipulated up and down for air.

Pottery vessels as artifacts needed external agents for movement. But the heart, as a natural vessel, by definition, moved itself.[208] By the elemental theory of directions, innate cardiac blood ascended by innate cardiac heat, despite the downward tendency of blood as earthy and watery. Aristotle compared the cardiac heat expanding its contents of blood to a boiling liquid, "for boiling is due to the volatilization of fluid by heat and the expansion consequent on increase of bulk."[209] The expansion caused pulsation as the blood rose toward its final membrane.[210] The pericardium was like a "dense and thin" skin enclosing the heart,[211] very large and strong to protect optimally the heart as the controller of life.[212] From his observation of the embryonic chick's heart in motion, Aristotle inferred that the heart, as the first organ, was the origin of all bodily movement.[213] That movement lay specifically in the sinews, the parts that held the animal together like glue.[214] Aristotle treated sinews, with bones, as formed by the solidification of fluid by heat. "Like earthenware (*keramos*), bones could not be shattered by fire; for as if in a kiln (*kaminos*) they were baked by fire in their formative heat."[215] But sinews were also tough. "In all animals their strength lies in their sinews," until slackened by age to become too weak or powerless to cause movement.[216] The starting point of the sinews, as of the blood vessels, was the heart.[217] "The heart also has many sinews, and this is reasonable. For the movements are from this part, and are accomplished through contracting and relaxing; so the heart needs such equipment and strength. As we said previously, the heart, in those that have it, is by nature like a sort of animal."[218] The ascent of the cardiac blood by heat was thus assisted by expanding cardiac sinews, much as an agent's grip on the *pnigeus* lifted it. The heart, as the origin of the sinews, had them in its largest cavity, the right one, to which the great blood vessel was attached.[219] The aorta was also "a sinewy blood-vessel" especially at its flexible ends.[220]

Aristotle compared the sinews to puppet strings, "the setting free and loosening of which causes the movement" of the figure, yet without alteration. "In the animal, however, the same part can become both greater and smaller and then again larger, and change its form, the members increasing through heat and contracting through cold and thus altering." Alterations around the heart caused visible bodily changes such as shivering from cold or its opposite.[221] Contraction and expansion were consequents of Aristotle's rejection in *Physica* of the void. They had to be possible or there could be no change, with loss or gain.[222] "Now the functions of movement are thrusting and pulling, so that the organ of movement must be able to increase and contract."[223] That ability applied primarily to the heart as the origin of all bodily movement. Aristotle thought that animal movement began from the right.[224] By inference, the movement of blood began from the sinewy right cavity of the heart to flow up into the great blood vessel and aorta. Because Aristotle varied the location of the aorta, its reception of the blood was inexact. If the aorta was attached to the left cavity, with the coolest blood, its own sinews compensated for the lesser energy of heat in that cavity. Although the aorta did not suck blood like a siphon, it had abundant sinews for movement at its flexible ends.

The oven *pnigeus* related to *pnigō*, "to throttle, strangle," and "to be of great heat." The Hippocratic Corpus knew both stifling atmospheric heat[225] and bodily suffocation in fevers.[226] In quinsy a cold and sticky defluxion from the head obstructed the passages of the breath and the blood, and it coagulated the blood flow. It caused convulsive suffocation[227] that stopped both the breath and the blood. Aristotle recorded that external pressure on the blood vessels along the windpipe caused choking, or else a shutting of the eyes and a fall.[228] *Rhetorica* included throttling as an analogy, about the Athenian general who was "strangling the state at the throat."[229] Aristotle was a strangler of animals, and not metaphorically but literally. He contrasted his account of the blood vessels with those of natural philosophers who had not described them as precisely. Their errors he attributed to "the difficulties of observation." He explained that in dead animals the most important blood vessels could not be examined because they had immediately collapsed, with their blood gushing out "as though from a vessel." For, the blood was not fluid in the body generally—only in the blood vessels and the heart. In live animals the blood vessels, being internal, were inaccessible to study.[230] Aristotle thus devised a method for their examination by dissection: starve, then strangle the animals. "It is only in strangled animals which have been previously emaciated that it is possible adequately to discover the facts, if one makes the subject one's business."[231] Aristotle made the blood flow his business, with the amphora as a cardiac model. Its tall slender neck had paired handles on its sides for grasping.[232] He may have thought of the heart, which was rounded at the top, with its protruding great blood vessel and the aorta as like a neck or handles of the vessel. An *aorta* meant a strap for hanging something, a loop that was graspable. The *pnigeus*, the domed oven, had no neck. But it had a handle on top for grasping to lift, set, or carry it. Just as the sinews of the human hand[233] deliberately grasped the *pnigeus* for those actions, the sinews of the heart naturally lifted and lowered it by expansion and contraction.[234] Pottery vessels usually had matching lids,[235] and transport amphoras had stoppers.[236] The heart had no lid or stopper to impede the blood flow—until death choked off its exits. For Aristotle, death in blooded animals was always by cardiac failure. By analogy with the amphora and that oven, death may have happened at the juncture between the heart and the great blood vessel and/or the aorta. With closure of the opening, blood stopped flowing out for the nourishment of the body.

Vases, it has been said, are more instructive about Greek antiquity in the fourth century than any other artistic medium.[237] They even depicted the lethal cardiac wounds of Homeric characters with anatomical precision.[238] Animals were also painted and sculpted, as hunted prey, sacrificial victims, domestic pets, and monstrous hybrids.[239] Aristotle preferred to study animals in nature rather than art, for he believed that the final cause with its beauty inhered more in nature. "It would be unreasonable, indeed absurd, to enjoy studying their representations on the grounds that we thereby study the art that fashioned them (painting or sculpture), but not to welcome still more the study of the actual things composed by nature, at least when we can survey their causes."[240] His criticism devalued his own anatomical drawings and, perhaps, also clay or wax models.[241] Yet, just as his dissections applied manual skill with the knife as a tool to examine hearts, he also resorted to man-made pottery for rational discovery and teaching. He modeled his cardiac vessel on the *amphoreus* for supplying and the *pnigeus* for cooking the dietary staples, the new wine and bread. Bread had long been basic food,[242] and by his century "wine had become part of daily life in virtually all of Greek society."[243]

Among Aristotle's contemporaries, physicians should have been ready to recognize his use of a cardiac vessel of blood as like their use of an amphora of wine. Physicians decanted wine as a common therapy. They prescribed wine extensively but discerningly for internal consumption and external application. As a drink, wine was a purgative and diuretic, expectorant and emetic. It was poured in surgery on wounds and fractures; it cleaned, injected, or fumigated the uterus. Wine made clysters, poultices, and ointments. And wine even substituted for blood when prescribed to compensate for loss from nosebleed and menstruation.[244] Aristotle regarded both the physician and the wine as agents of healing, with the art of medicine as the first mover and the wine as the last mover because it was moved by the physician.[245] He praised the skill of physicians in dispensing wine for health as equal to their professional knowledge. He was realistic, not idealistic, about medicine. Physicians, as physicians, did not need to know the ideal Good or even to study abstract health. They needed to understand how to cure the particular humans in their care.[246] It was Aristotle's own ingenuity in reason and research to exploit the ordinary to investigate the extraordinary, the hidden hearts of animals from human artifacts in plain sight.

Notes

1. Aristotle, *De partibus animalium* 1.5 645a; trans., A. L. Peck, *Parts of Animals* (Cambridge, Mass.: Harvard University Press, 1961), p. 101.
2. *De partibus animalium* 2.3 650b; 3.4 665b, 666a, 667a.
3. Ibid. 666a, 666b; trans. James G. Lennox, *Aristotle "On the Parts of Animals"* (Oxford: Clarendon, 2001), pp. 56, 57. This translation is usual hereafter.
4. James Rochester Shaw, "Models for Cardiac Structure and Function in Aristotle," *Journal for the History of Biology* 5 (1972): 378, 385.
5. See C. R. S. Harris, *The Heart and the Vascular System in Ancient Greek Medicine, from Alcmaeon to Galen* (Oxford: Clarendon, 1973), pp. 121–76; Shaw, "Models," pp. 355–88.
6. John Boardman, *The History of Greek Vases: Potters, Painters, and Pictures* (London: Thames & Hudson, 2001), pp. 9, 244. For usage, see Brian A. Sparkes, *Greek Pottery: An Introduction* (Manchester: Manchester University Press, 1991), pp. 60–65, 71–92.

7. For an introduction, see Gisela M. Richter and Marjorie J. Milne, *Shapes and Names of Athenian Vases* (New York: Metropolitan Museum of Art, 1935). For an introduction to the classifications, see also The Beazley Archive, University of Oxford: www.beazley.ox.ac.uk.
8. Critias ap. Athenaeus, *Deipnosophistai* 1.28b; trans. Kathleen Freeman, *Ancilla to the Pre-Socratic Philosophers* (Oxford: Basil Blackwell, 1956), p. 154. Cited by François Lissarrague, *The Aesthetics of the Greek Banquet: Images of Wine and Ritual*, trans. Andrew Szegedy-Maszak (Princeton: Princeton University Press, 1990), p. 140; Brian A. Sparkes and Lucy Talbott, *Pots and Pans of Classical Athens* (Princeton, N. J.: American School of Classical Studies at Athens, 1958), p. 2. For Aristotle's rejection of Critias on perceptive blood and the soul, see Aristotle, *De anima* 1.2 405b; *De partibus animalium* 3.4 666a.
9. John W. Hoopes and William K. Barnett, "The Shape of Early Pottery Studies," in *The Emergence of Pottery: Technology and Innovation in Ancient Societies*, ed. idem (Washington, D.C.: Smithsonian Institute Press, 1995), pp. 1–7.
10. Eleni Hasaki, "Craft Apprenticeship in Ancient Greece: Reading Beyond the Masters," in *Archaeology and Apprenticeship: Body, Knowledge, Identity, and Communities of Practice*, ed. Willeke Wendrich (Tucson: University of Arizona Press, 2012), p. 258.
11. Lisa C. Kahn and John C. Wissinger, "Recreating and Firing a Greek Kiln," in *Papers on Special Techniques in Athenian Vases*, ed. Kenneth Lapatin (Los Angeles, Calif.: The J. Paul Getty Museum, 2008), pp. 129–38. For the updraught kiln, see also J. M. Hemelrijk, "A Closer Look at the Potter," in *Looking at Greek Vases*, ed. Tom Rasmussen and Nigel Spivey (Cambridge: Cambridge University Press, 1991), pp. 243–44.
12. Boardman, *History*, p. 165; John Boardman, *Athenian Red Figured Vases: The Classical Period: A Handbook* (London: Thames & Hudson, 1989), p. 234; Andrew J. Clark, Maya Elston, and Mary Louise Hart, *Understanding Greek Vases: A Guide to Terms, Styles and Techniques* (Los Angeles, Calif.: The J. Paul Getty Museum, 2002), pp. 3, 23.
13. Diogenes Laertius, *Vitae philosophorum* 5.22–27. See also Paul Moraux, *Les listes anciennes des ouvrages d'Aristote* (Louvain: Éditions universitaires de Louvain, 1951), pp. 15–193.
14. Carlo Natali, *Aristotle: His Life and School*, ed. D. S. Hutchinson (Princeton: Princeton University Press, 2013), pp. 96–104, 148–49, 150. See also James G. Lennox, "The Disappearance of Aristotle's Biology: A Hellenistic Mystery," in idem, *Aristotle's Philosophy of Biology: Studies in the Origins of Life Science* (Cambridge: Cambridge University Press, 2001), pp. 110–25.
15. See Natali, *Aristotle*, pp. 115–16.

16. Aristotle, *De generatione animalium* 2.6 743b; trans. A. L. Peck, *Generation of Animals* (Cambridge, Mass.: Harvard University Press, 1943), p. 225.
17. Aristotle, *Poetica* 2.6 1450a.
18. See below page 12.
19. Clark et al., *Understanding*, p. 1.
20. Plato, *Politicus* 287d–288a; trans. Christopher J. Rowe, *Statesman* (Warminster: Aris & Phillips, 1995), p. 111.
21. Aristotle, *Physica* 2.1 192b and see 1.8 191b; trans. Philip H. Wicksteed and Francis M. Cornford, 2 vols. (Cambridge, Mass.: Harvard University Press, 1980), 1:107–9. See also Aristotle *De generatione animalium* 2.1 734b–735a. Cf. Aristotle, *Metaphysica* 3.5 1002a. For knowledge and craft in general, see Aristotle, *Ethica nicomachea* 6.
22. Boardman, *Classical*, p. 234; Dyfri Williams, *Greek Vases*, 2nd ed. rev. (London: British Museum Publications, 1999), p. 138.
23. Boardman, *History*, p. 153; Hasaki, "Apprenticeship," p. 178.
24. John Boardman, *Athenian Red Figured Vases: The Archaic Period: A Handbook* (London: Thames & Hudson, 1975), p. 7.
25. Clark et al., *Understanding*, p. 11. For style, see Williams, *Vases*, pp. 112–30; Hemelrijk, "Look," p. 256; Boardman, *Classical*, pp. 7, 190.
26. Aristotle, *Rhetorica* 1.2.4 1381b; Aristotle, *Politica* 5.10 1312b5; *Ethica nicomachea* 7.1.7 1235a.
27. Boardman, *History*, p. 143.
28. Lissarrague, *Aesthetics*, p. 95; Boardman, *History*, p. 262.
29. Aristotle, *Historia animalium* 6.2 590a.
30. Ibid. 6.2 559b.
31. *De generatione animalium* 3.2 753a. For amphoras, see Raymond Billiard, *La vigne dans l'antiquité* (Marseille: Laffitte Reprints, 1997), pp. 514–21, and for spoiling, pp. 532–35. For wine and vinegar, see *Metaphysica* 8 1044b.
32. *Historia animalium* 4.18 549b–550a; 4.1 525a. For generation of oysters in vessels, see *De generatione animalium* 3.11 763a.
33. *Historia animalium* 4.1 525a.
34. Ibid. 6.37 580b. For hibernation of birds in vessels, see 6.16 600a.
35. John M. Camp, *The Archaeology of Athens* (New Haven, Conn.: Yale University Press, 2001), pp. 261–64; Boardman, *Classical*, p. 234; idem, *History*, p. 139 and see p. 141. For the workshops and processes in general, see Sparkes, *Pottery*, pp. 10–26. For a scene of an apprentice turning the wheel, see Sparkes and Talbott, *Pots and Pans*, fig. 1.
36. Without Aristotle, see Camp, *Archaeology*, p. 263; Ada Caruso, *Akademia: Archeologia de una scuola filosofica ad Atene da Platone a Proclo (387 a.c.–485 d.c)* (Athens: Scuola archeologica italiana di Atene, 2013). For Aristotle at the Academy, see Natali, *Aristotle*, pp. 17–31.

37. Hasaki, "Apprenticeship," pp. 187–88, citing Plato, *Respublica* 5 467a, trans. Benjamin Jowett; Sparkes, *Pottery*, p. 10.
38. Joseph V. Noble, *The Techniques of Attic Painted Pottery* (New York: Watson-Guptill with the Metropolitan Museum of Art, 1965), pp. 5–6, 15–16. For Aristotle on the pithos, see *Physica* 4.6 213b.
39. See Sparkes, *Pottery*, pp. 10–26; Eleni Hasaki, "Workshops and Technology," in *A Companion to Greek Art*, ed. Tyler Jo Smith and Dimitris Plantzos, 2 vols. (Oxford: Basil Blackwell, 2012), 1: 257–62.
40. Aristotle, *Protrepticus*, under reconstruction (September 2017 version) @ www.protrepticus.info, p. 20. See also Douglas Hutchinson and Monte R. Johnson, "Authenticating Aristotle's *Protrepticus*," *Oxford Studies in Ancient Philosophy* 29 (2005):193–294.
41. Principal texts are Aristotle, *Meteorologica* 4, and Aristotle, *De generatione et corruptione* 1.1 314a–315b; 1.6 322b–323a; 1.10 327a–328b. For animals, see *De partibus animalium* 2.1 646a. See also G. E. R. Lloyd, *Aristotle: The Growth and Structure of His Thought* (Cambridge: Cambridge University Press, 1968), pp. 160, 162, 164–75; James Longrigg, *Greek Rational Medicine: Philosophy and Medicine from Alcmaeon to the Alexandrians* (London: Routledge, 1993), pp. 151–59; David J. Furley, "The Mechanics of *Meteorologica* IV: A Prolegomenon to Biology," in idem, *Cosmic Problems: Essays on Greek and Roman Philosophy of Nature* (Cambridge: Cambridge University Press, 1989), pp. 132–48.
42. *Meteorologica* 4.5 382a; trans. H. D. P. Lee, *Meteorologica* (Cambridge, Mass.: Harvard University Press, 1978), p. 315.
43. Noble, *Techniques*, pp. 1–2; Toby Schreiber, *Athenian Vase Construction: A Potter's Analysis* (Malibu, Calif.: The J. Paul Getty Museum, 1999), pp. 3–8; Williams, *Vases*, p. 11; Sparkes, *Pottery*, pp. 8–10; Hemelrijk, "Look," p. 236.
44. *Meteorologica* 4.7 384b.
45. Ibid. 4.5 382a–384b. See also *De partibus animalium* 2.2 649a.
46. *Meteorologica* 4.7 384a, b; *De partibus animalium* 2.4 650b, 3.5 668b.
47. *Meteorologica* 4.6 383a; trans., p. 321. This analysis is not among the recognized flaws and defects of the classical Greek kiln. Without mention of Aristotle, see Noble, *Techniques*, pp. 79–83; Schreiber, *Construction*, pp. 56–66; Sparkes, *Pottery*, pp. 15, 24, 25. For Attic clay as "tough," see Schreiber, *Construction*, pp. 5, 37.
48. *De generatione animalium* 2.6 743a; trans., pp. 219–21.
49. *Meteorologica* 4.7 384b; trans., p. 337.
50. Hemelrijk, "Look," p. 236; Boardman, *History*, p. 244.
51. Schreiber, *Construction*, pp. xiii, 53; Sparkes, *Pottery*, p. 15.
52. Carolyn G. Koehler, "Handling of Greek Transport Amphoras," in *Recherches sur les amphores grecques*, ed. J.-Y. Empereur and Y. Garlan (Athens: École francaise d'Athens, 1986), pp. 50–52.

53. *De partibus animalium* 3.4 665b–666a; trans., p. 56.
54. *Historia animalium* 3.20 521b; trans., 1:223. ibid. 3.19 520b, trans. mine. Cf. 1.17 496b; 1.34 489a, 3.2 511b.
55. *A Greek English Lexicon*, ed. Henry G. Liddell and Robert Scott (Oxford: Clarendon, 1968) s.v. All further references to Greek words are to this edition.
56. Vivian Nutton, *Ancient Medicine*, 2nd ed. (London: Routledge, 2012), p. 45.
57. *On the Heart* is dated very tentatively to the 270s by Nutton, *Ancient Medicine*, p. 139.
58. *Regimen in Acute Diseases* 21, ed. W. H. S. Jones, in *Hippocrates*, vol. 2.
59. *Barrenness* 12 432, ed. Paul Potter, in *Hippocrates*, vol. 10.
60. *Diseases* 4.6 554, ibid.
61. For *De partibus animalium* as prior to *Historia animalium*, see D. M. Balme with Allan Gotthelf, ed., *History of Animals Books VII–X* (Cambridge, Mass.: Harvard University Press, 1991), pp. 21–23.
62. *Protrepticus*, p. 53.
63. Aristotle, *De longitudine vitae, de iuventute* 27 480b, in *Parva naturalia*. This edition does separate the chapters on respiration.
64. Nutton, *Ancient Medicine*, p. 119. See also G. E. R. Lloyd, *In the Grip of Disease: Studies in the Greek Imagination* (Oxford: Oxford University Press, 2003), pp. 176–79; Longrigg, *Medicine*, pp. 150–51. For Aristotle's reputed descent from Asclepiad physicians and for his reputed practice of medicine, see Natali, *Aristotle*, pp. 8–11.
65. Aristotle, *Categoriae* 15b; trans. J. L. Ackrill, *Categories and De interpretatione* (Oxford: Clarendon, 1963), p. 42.
66. Aristotle, *Metaphysica* 5.23 1023a, as noted by Ackrill, ed., *Categories*, p. 112. *Metaphysica*, trans., p. 277.
67. For a philosophical analysis, see Benjamin Morison, *On Location: Aristotle's Concept of Place* (Oxford: Clarendon, 2002), pp. 54–80, 121–32.
68. *Physica* 4.1 208b.
69. Ibid. 4.2 209b; trans., p. 291.
70. Ibid. 4.3 210a; trans., p. 295
71. Ibid. 4.3 210a; trans., p. 295.
72. Ibid. 4.3 210b; trans., p. 299.
73. *Metaphysica* 8.2, 1043a.
74. *Politica* 1.4 1253b–1254a.
75. Nicholas Cahill, *Household and City Organization at Olynthus* (New Haven, Conn.: Yale University Press, 2002), pp. 169, 226.
76. *Physica* 4.6 213b.
77. Cahill, *Organization*, pp. 227–28.

78. Noel Robertson, "Athens' Festival of the New Wine," *Harvard Studies in Classical Philology* 95 (1993):197–250.
79. See Koehler, "Handling," p. 58.
80. Cahill, *Organization*, p. 229.
81. Schreiber, *Construction*, p. 73.
82. Noble, *Techniques*, pp. 17–18; Schreiber, *Construction*, pp. 128–43. For Aristotle on mixing wine with water, see *De generatione et corruptione* 1.5 321a–b, 322a.
83. Virginia Grace, "Standard Pottery Containers of the Ancient Greek World," *Hesperia Supplements* 8 (1949), pp. 175–76.
84. *Historia animalium* 3.20 522a.
85. *Ethica nicomachea* 8.2 1155a; 7.14 1154a, and see 3.9–10 1118a.
86. *Historia animalium* 3.19 520b, 521a; *De iuventute et senectute* 20 480a.
87. *De iuventute et senetute* 20 480a.
88. *De partibus animalium* 2.1 647b; trans., p. 18.
89. *De iuventute et senectute* 20 480a; trans. W. S. Hett, *"On the Soul," "Parva naturalia," "On Breath"* (Cambridge, Mass.: Harvard University Press, 1986), p. 477.
90. See *Physica* 8.7 260a.
91. *Historia animalium* 3.20 521b.
92. *De partibus animalium* 2.2 647b.
93. Ibid. 2.3 650b; trans., p. 24.
94. Aristotle, *De caelo* 3.7 305b; see also *De partibus animalium* 1.1 640b.
95. *De partibus animalium* 3.4 665b; 666b; trans., p. 57; 666a trans., p. 56; *De generatione animalium* 2.5 741b; *De iuventute et senectute* 3 468b–469a, 8 474b; *De generatione animalium* 2.4 740b.
96. *De partibus animalium* 2.1 647b, 3.4 665b.
97. Ibid. 2.3 650b; trans., p. 24.
98. Ibid. 2.3 650a, 3.4 667a, 3.5 667b; *Historia animalium* 3.2 511b; *De generatione animalium* 2.4 740a.
99. *De partibus animalium* 4.5 667b; trans., p. 59.
100. Ibid. 3.4 665b; trans., p. 55. See also 2.7 652a, b.
101. Ibid. 3.4 666a; trans., p. 56. *De iuventute et senectute* 3 468b–469a, 8 474b.
102. *Politica* 5.1 1301b.
103. See Hemelrijk, "Look," p. 234; Clark et al., *Understanding*, p. 4; Sparkes, *Pottery*, pp. 78–79.
104. Lissarrague, *Aesthetics*, pp. 56–58.
105. Christiane Bron and François Lissarrague, "Looking at the Vase," in *A City of Images: Iconography and Society in Ancient Greece*, ed. Claude Bérand and Bron, trans. Deborah Lyons (Princeton, N.J.: Princeton University Press, 1989), pp. 12–14; Sparkes, *Pottery*, p. 79.

106. *De generatione animalium* 1.22 730b; trans., p. 119.
107. For vessels of air for speaking and hearing, see ibid., 5.7 787b; *Historia animalium* 1.11 492a.
108. *De partibus animalium* 3.8 671a; 4.1 676a; and see 4.1 676a, 4.5 680b; 3.8 670b–671a; 671b.
109. *De generatione animalium* 4.8 776b; trans., p. 469.
110. *De partibus animalium* 4.10 688a; *De generatione animalium* 3.2 752b, 4.8 776a.
111. *De generatione animalium* 2.4 739b, 4.8 776a.
112. *Meteorologica* 4.5 382b, 4.7 384a.
113. *De partibus animalium* 4.10 688a–b; *Historia animalium* 2.1 500a.
114. *De generatione animalium* 4.8 776a; trans., p. 469.
115. *De partibus animalium* 4.10 688a.
116. Ibid. 3.4 665a.
117. Ibid. 4.11 692a; and see 1.1 640b.
118. Ibid. 3.3 664b; 4.5 678b, 4.6 682a; 3.14 675b, 4.6 682a.
119. Ibid. 3.4 666a.
120. Ibid. 3.4 666b; trans., p. 57; *Historia animalium* 1.17 496a, 3.3 513a.
121. *De partibus animalium* 3.4 665b–666a.
122. Ibid. 3.4 666b–667a; *Historia animalium* 1.17 496a, 3.3 513a. For a philosophical analysis, see Lennox, trans., pp. 259–60.
123. Ibid. 3.4 667a. A common vessel within a vessel was the wine cooler (*psykter*). Clark et al., *Understanding*, p. 134; Schreiber, *Construction*, pp. 218–23; Noble, *Techniques*, p. 19; Sparkes and Talbott, *Pots and Pans*, figs. 18, 19.
124. Aristotle, *Topica* 6.4 141a, b.
125. *De partibus animalium* 3.4 665b; trans., p. 55; *Historia animalium* 6.3 561a, 562a.
126. *De partibus animalium* 3.4 666a.
127. *De generatione animalium* 3.3 754b; trans., p. 299.
128. *Historia animalium* 6.3 561a; trans., 2:235.
129. *De partibus animalium* 3.4 666b; trans., p. 57.
130. *Historia animalium* 1.17 496b.
131. *Physica* 2.2, 193b–194a.
132. Boardman, *History*, pp. 262, 423.
133. Schreiber, *Construction*, pp. 73–80; Noble, *Techniques*, pp. 12–14; Grace, "Containers," pp. 175–89, 455–56; Boardman, *History*, p. 282; Clark et al., *Understanding*, pp. 66–67.
134. Grace, "Containers," p. 175 and pl. 19; Noble, *Techniques*, fig. 85, p. 144; Boardman, *History*, p. 262.
135. See Hemelrijk, "Look," p. 234.
136. Noble, *Techniques*, p. 13.

137. Virginia Grace, *Amphoras and the Ancient Wine Trade* (Princeton, N.J.: American School of Classical Studies at Athens, 1961), p. 9; idem, "Containers," p. 175; Koehler, "Handling," p. 60.
138. Koehler, "Handling" pp. 61–62; Noble, *Techniques*, p. 13; Grace, *Amphoras*, pp. 2–3.
139. Koehler, "Handling." p. 58; Sparkes and Talbott, *Pots and Pans*, fig. 16.
140. Grace, *Amphoras*; Koehler, "Handling," p. 58.
141. For trade, see Boardman, *History*, pp. 153–67; Sparkes, *Pottery*, pp. 131–35.
142. *De partibus animalium* 3.5 667b; trans., p. 59.
143. *Historia animalium* 3.3 513b; *De partibus animalium* 3.4 667a.
144. Aristotle, *De somno et vigilia* 3 458a, in *Parva naturalia*; *De partibus animalium* 3.4 667a.
145. *De partibus animalium* 3.4 667a; *Historia animalium* 1.17 496a, 3.3 513b.
146. Aristotle, *Incessu animalium* 2 704b 18–22; 4 705a–b, 5 706b; 6. 708a. *Historia animalium* 1.14 493b; 1.15 494a; *De partibus animalium* 2.2 648a. For animal movement, see also Martha Nussbaum, ed., *Aristotle's "De motu animalium"* (Princeton, N. J.: Princeton University Press, 1978), pp. 107–42. *Physica* 4.1 208b, and see G. E. R. Lloyd, "Right and Left in Greek Philosophy," in idem, *Methods and Problems in Greek Science* (Cambridge, Mass.: Cambridge University Press, 1991), pp. 41–46.
147. *De partibus animalium* 3.4 665b; trans., p. 55.
148. *Physica* 4.1 208b; trans., p. 279.
149. *Meteorologia* 4.7 384a, b.
150. Grace, *Amphoras*, p. 3 and fig. 7; Lissarrague, *Aesthetics*, p. 23 and fig. 10; Sparkes and Talbott, *Pots and* Pans, fig. 14; Richter and Milne, *Shapes*, pp. 6–7.
151. Sparkes, *Pottery*, p. 61; Grace, *Amphoras*, pp. 1–2. For ladle or pitcher, see Clark et al., *Understanding*, pp. 118–19; Sparkes and Talbott, *Pots and Pans*, fig. 17.
152. Koehler, "Handling," p. 66.
153. Citing Hippon. Cf. Aristotle, *Metaphysica* 1.3 984a; *De anima* 1.2 405b. The practice dated to the fourteenth century B.C.E., as attested on a Syrian gravestone. Grace, *Amphoras*, p. 3 and fig. 8.
154. Alexander of Aphrodesias *Problemata* 2.59, cited at Aristophanes, *Thesmorphoriazusae* line 557, by eds. Colin Austin and Douglas Olsen (Oxford: Oxford University Press, 2004), p. 216.
155. Sylvia Berryman, *The Mechanical Hypothesis in Ancient Greek Natural Philosophy* (Cambridge: Cambridge University Press, 2009), pp. 55, 58, 232, and see pp. 97–103.
156. *De partibus animalium* 3.4 666a; trans., p. 56.

157. *Meteorologica* 2.1 353b; trans. E. W. Webster, in *Works*, 12 vols. (Oxford: Clarendon, 1952–63), 3:590.
158. *De partibus animalium* 3.5 668a.
159. For Klepsydra but without mention of Aristotle, see Dora P. Crouch, *Water Management in Ancient Greek Cities* (New York: Oxford University Press, 1993), pp. 262–63, citing Arthur W. Parsons, "Klepsydra and the Paved Court of the Pythion," *Hesperia* 12 (1943): 192. For that spring as "never dry," "unusually constant," see Parsons, pp. 203, 223. For the construction of the spring house for the drain basin, see pp. 208–22. Building a house from stone as being like the blood vessels was Aristotle's immediate parallel to them as watercourses in gardens. *De partibus animalium* 3.5 668a. The actual spring and house on the Acropolis recommends Aristotle's disparate analogies as coherent. For the heart as an acropolis of the body, see ibid. 3.7 670a and below.
160. *De partibus animalium* 3.4 667b; trans., p. 59.
161. Ibid. 3.4 666a; trans., p. 56.
162. *Ethica nicomachea* 5.9 1137a; trans. Robert C. Bartlett and Susan D. Collins, *Aristotle's "Nicomachean Ethics"* (Chicago: University of Chicago Press, 2012), pp. 110–11.
163. *Protrepticus*, p. 17.
164. *Meteorologica* 4.1 378b; trans., Lee, p. 291. See also 4.4–7; *De partibus animalium* 2.3 648a–649b.
165. *Meteorologica*. 2.9 369a, b; trans., pp. 225, 227. 1.2 339a.
166. *De partibus animalium* 3.4 667b; cf. *De generatione animalium* 2.3 737a. The vital heat and the issues of pneuma and soul involve philosophical inquiries beyond the scope of this study. For an introduction, see Gad Freudenthal, *Aristotle's Theory of Material Substance: Heart and Pneuma, Form and Soul* (Oxford: Clarendon, 1995). See also forthcoming, *Aristotle and His Predecessors on Heat, Pneuma, and Soul*, ed. Hynek Bartoš and Colin Guthrie King (Cambridge: Cambridge University Press), in press from a conference in Prague in June 2014.
167. *De partibus animalium* 3.5 667b.
168. *Meteorologica* 4.7 384a, 4.11 389b; *De partibus animalium* 2.4 650a, 3.5 668a.
169. *De partibus animalium* 3.4 665b–666a; trans., p. 56.
170. Ibid. 3.7 670a; trans., p. 64. For the Acropolis in Athens as suited to oligarchy or monarchy but not democracy, see *Politica* 7.11 1330b.
171. Mika Kajava, "Hestia Hearth, Goddess, and Cult," *Harvard Studies in Classical Philology* 102 (2004):1.
172. Barbara Tsakirgis, "Fire and Smoke: Hearths, Braziers, and Chimneys in the Greek House," *British School at Athens Studies* 15 (2007): 226–28.

173. Angela Della Volpe, "From the Hearth to the Creation of Boundaries," *Journal of Indo-European Studies* 18 (1990):158–60; Tsakirgis, "Fire and Smoke," p. 230.
174. *De motu animalium* 10 703a; *Politica* 1.2 1253a.
175. *Politica* 6.8 1322b. For the communal hearth, see Lin Foxhall, "House Clearance: Unpacking the 'Kitchen' in Classical Greece," *British School at Athens Studies* 15 (2007): 234.
176. *De partibus animalium* 3.4 667a.
177. *De anima* 1.2 405a; trans. J. A. Smith, in *Works*, 3:206–7.
178. For previous interpretations, see Pavel Gregoric, "The Heraclitus Anecdote: *De partibus animalium* i 5.645a12–23," *Ancient Philosophy* 21 (2001): 73–85. His interpretation based on the "kitchen" in "private women's quarters," on pp. 79, 83 did not consider Lisa C. Nevitt, "Gender Relations in the Greek Household: The Archaeological Evidence," *Annual of the British School at Athens* 90 (1995): 363–81. For accuracy, see also Tsakirgis, "Fire and Smoke"; Foxhall, "House Clearance," pp. 233–43.
179. See Sparkes, "Kitchen," pp. 129, 127. For the unusual "kitchen-complex" at Olynthus, see Cahill, *Organization*, pp. 153–56, 161–69.
180. *De anima* 1.2 404b; trans., p. 204.
181. *De partibus animalium* 3.4 667a; trans., p. 58.
182. Aristotle, *Poetica* 2. 7 1450b–1451a; trans. Anthony Kenny, *Poetics* (Oxford: Oxford University Press, 2013), p. 26.
183. *De partibus animalium* 1.5 645a; trans., p. 14.
184. Ibid. 1.5 645a.
185. *Historia animalium* 5.32 557b; trans., 2:215.
186. *De partibus animalium* 3.4 665a.
187. *De iuventute et senectute* 8 474a–b.
188. See G. E. R. Lloyd, "The Master Cook," in *Aristotelian Explorations* (Cambridge: Cambridge University Press, 1996), pp. 83–103.
189. *Meteorologica* 4.2 379b–4.3 381b.
190. Ibid. 4.10 388b; 4.7 384a, 4.11 387b; cf. 4.9 387b.
191. Ibid. 4.5 382b; 4.7 384a. See also *De partibus animalium* 2.7 653a; *De generatione animalium* 2.6 743b. For the process, see Paul Lukacs, *Inventing Wine: A New History of One of the World's Most Ancient Pleasures* (New York: W. W. Norton, 2012), pp. xiii, 20, 21; Billiard, *Vigne*, pp. 497–99; Brian A. Sparkes, "Treading the Grapes," *Babesch* 51 (1976): 47.
192. *Meteorologica* 4.7 384a, b; *Historia animalium* 3.6 515b–516a.
193. See Hemelrijk, "Look," p. 237; Boardman, *History*, p. 266; Noble, *Techniques*, pp. 5–6; Brian A. Sparkes, "The Greek Kitchen (Plates IV–VIII),"

Journal of Hellenic Studies 82 (1962): 124 [121–37], and see idem, "The Greek Kitchen: Addenda," ibid. 85 (1965): 162–63.
194. Sparkes, "Kitchen," p. 127 and fig. 1; p. 128 fig. 2; Pls. VIII 2, VIII 4.
195. *De partibus animalium* 1.5 645a.
196. Sparkes and Talbott, *Pots and Pans*, fig. 36; Sparkes, "Kitchen," p. 128 and pl. IV 2.
197. Aristophanes, *Nubes* lines 94–96; trans. William J. Hickie, *The Clouds* (New York: Hinds & Noble, n.d.), p. 7. See also *Aves* line 1001.
198. *De partibus animalium* 2.8 654a; trans. mine.
199. *De iuventute et senectute* 17 479a.
200. Ibid. 4–5 469b, 8 474a–b; 17 478b, trans., p. 471; 17 479a, 5 469b–470a, 8 474b, 14 474a–b, 17 479a.
201. Ibid. 5 470a.
202. Personal communication from Brian A. Sparkes to verify his photographs.
203. *De longitudine et brevitate vitae* 3 465b, 5 466b; *De iuventute et senectute* 5 470a, 17 479a.
204. *De iuventute et senectute* 15 478a.
205. See Boardman, *History*, p. 140 pl. 173.2.
206. *De partibus animalium* 3.6 668b–669a; *Historia animalium* 496a; *De iuventute et senectute* 10 475b, 13 477a15 478a–16 477a. For the brain cooling, see *De partibus animalium* 2.7 652b.
207. *De iuventute et senectute* 16 478b; and like a bellows, 7 474a, 21 480a–b.
208. *Physica* 2.1 192b.
209. *De iuventute et senectute* 20 479b; trans. G. R. T. Ross, in *Works*, 3:432.
210. Ibid. 20 480a.
211. *Historia animalium* 3.13 519a–b.
212. *De partibus animalium* 3.11 673b.
213. Ibid. 3.4 666a.
214. *De generatione animalium* 2.3 737b.
215. Ibid. 2.6 743a; trans. mine. Cf. 2.3 737a.
216. Ibid. 5.7 787b.
217. *Historia animalium* 3.5 515a.
218. *De partibus animalium* 3.4 666b; trans., p. 57. Cf. 4.13 696.
219. *Historia animalium* 1.17 496a, 3.3 513a, 3.5 515a; *De partibus animalium* 3.4 666b–667a.
220. *Historia animalium* 3.5 515a, b; 3.4 514b.
221. *De motu animalium* 7 701b; trans. E. S. Forster, in *Parts of Animals, Movement of Animals, Progression of* Animals, ed. A. L. Peck and Forster (Cambridge, Mass.: Harvard University Press, 1961), p. 465.
222. *Physica* 4.9 213a–216b.

223. *De motu animalium* 10 703a; trans., p. 475.
224. *Incessu animalium* 4 705b–706a.
225. *On Ancient Medicine* 16; *On Airs, Waters, and Places* 10, ed. Jones, in *Hippocrates*, vol. 1.
226. *Aphorisms* 4.34, 7.59; *Epidemics* 3.3; *Prognostics* 23; *On the Sacred Disease* 1, ed. Jones, in *Hippocrates*, vol. 2.
227. *On Regimen in Acute Diseases* appendix 6, ed. Paul Potter, *Hippocrates*, vol. 6, p. 6.
228. *Historia animalium* 3.3 514a. See also *De iuventute et senectute* 9 475a.
229. *Rhetorica* 3.10 1411a; trans. George A. Kennedy, *On Rhetoric: A Theory of Civic Discourse*, 2nd ed. (New York: Oxford University Press, 2007), p. 220.
230. *Historia animalium* 3.2 511b; trans., 1:163.
231. Ibid. 3.3 513a; trans., 1:173.
232. For handles, see Schreiber, *Construction*, pp. 24–26, 36–37, 79–80 and *passim*; Noble, *Techniques*, figs. 92–105 and *passim*.
233. *De partibus animalium* 2.1 646b.
234. See *De generatione animalium* 5.7 787b.
235. See Noble, *Techniques*, pp. 4, 13; Schreiber, *Construction*, pp. 23, 36, 80, 262.
236. Grace, *Amphoras*, p. 1; Koehler, "Handling," pp. 52–54, 56.
237. Boardman, *History*, p. 7.
238. S. Garoulanos, A. Tasouli, E. Lymberopoulou, and K. Papadoulos, "Homeric Injury Scenes on Ancient Greek Pottery Reveal Medical Knowledge," in *Science and Technology in Homeric Epics*, ed. S. A. Paipetis (New York: Springer, 2009), p. 272 fig. 3, p. 273 fig. 4. See also David Saunders, "Warriors' Injuries on Red-Figure Vases," *Mouseion* 10 (2010): 1–21.
239. Judith Barringer, *The Hunt in Ancient Greece* (Baltimore, Md.: Johns Hopkins University Press, 2001); Folkert T. Van Straten, *"Hiera kala": Images of Animal Sacrifice in Archaic and Classical Greece* (Leiden: E. J. Brill, 1995); Kenneth Kitchell, "Penelope's Geese: Pets of the Ancient Greeks," *Expedition* 53 (2011): 14–23; J. Michael Padgett, *The Centaur's Smile: The Human Animal in Early Greek Art* (Princeton: Princeton University Press, 2003).
240. *De partibus animalium* 1.1 639a; 1.5 645a; trans. D. M. Balme, *Aristotle's "De partibus animalium" I and "De generatione animalium" I (with passages from II.1–3)* (Oxford: Clarendon, 1972), p. 18. For representation vs. facts of animals, see *Poetica* 4.25 1460. For inborn pleasure in representations, see 1.4 1448b.
241. See *Historia animalium* 3.5 515b; *De generatione animalium* 2.6 743a, 4.1 764b.

242. For diet, see Sparkes, "Kitchen," p. 123; James Davidson, "*Opsophagia*: Revolutionary Eating at Athens," in *Food in Antiquity*, ed. John Wilkins, David Harvey, and Mike Dobson (Exeter: University of Exeter Press, 1995), p. 205; Robertson, "Festival," p. 197.
243. Lukacs, *Wine*, p. 16.
244. Jacques Jouanna, "Le vin et la medicine dans la Grèce antique," in *Greek Medicine from Hippocrates to Galen: Selected Papers*, ed. Philip van der Eijk, trans. Neil Allies (Leiden: Brill, 2012), pp. 173, 179, 176, 186–92, 184. See also Robert Alessi, "Le vin dans les *Épidémies* d'Hippocrate," in *Vin et santé en Grèce ancienne*, ed. Jacques Jouanna and Laurence Villard with Daniel Béguin (Paris: École française d'Athenes, 2002), pp. 105–12.
245. *De generatione et corruptione* 1.7 324a–b.
246. *Ethica nicomachea* 5.9 1137a, 1.6 1097a.

Augustine's Law of the Heart: Thieves' Honor

Augustine's adolescent theft of pears is a perplexing episode of his *Confessions*, his personal epideictic rhetoric in praise of God and blame of self. It confesses how he and his companions prolonged their nocturnal sporting at a farm near his family's rural estate. There they picked the neighbor's ripe pears but, after tasting a few in disgust, tossed loads to pigs.[1] Augustine's serious repentance and analysis for what seems a venial sin of pilfering has motivated formal investigation of his ethics.[2] Scholarly interpretation compares that pear tree near Thagaste in Roman North Africa with the fruit tree in the biblical garden of Eden from which Adam ate as the original sin. Augustine thus symbolically acknowledged himself an heir to Adam's fallen concupiscence.[3] Yet, Adam did not initiate that deed; his helper, Eve, did. "She took of its fruit and ate; and she also gave some to her husband, and he ate."[4] Neither does Augustine initiate his theft of pears; his crowd does, and that complicity is significant. Augustine's examination of conscience deliberately elaborates on their collaboration. As he confesses succinctly, "But when it is said, 'Let's go, let's do it,' we are ashamed not to be shameless."[5] Beyond philosophical and theological interpretations of his theft, Augustine's indictment of it belongs historically in the context of his social and cultural environments. His *Confessions* blames his collaboration in the theft on the exemplary rhetoric of the Roman comedic theater to prompt the applause of his peers for his sexual play. Augustine acknowledges the punishment of theft not only by "thy

M. O'R. Boyle, *Cultural Anatomies of the Heart in Aristotle, Augustine, Aquinas, Calvin, and Harvey*,
https://doi.org/10.1007/978-3-319-93653-6_2

law," meaning biblical law, but also by "the law written in human hearts." Augustine's law of the heart is customary behavior from social sympathy: like likes like and acts the same. In his hyperbole that "law" is outlawry, "for what thief suffers a thief with an even mind."[6]

Laws

Augustine prefaces the episode of the theft with his purpose in confessing it. "I intend to recollect my foul affairs and the carnal corruptions of my soul, not so that I might love them, but so that I might love thee, my God. I compose that subject with love for thy love, recalling my gravely wanton ways in the bitterness of my reconsideration."[7] His intention conforms to the theme of a personal encomium, the rhetorical portrayal of an individual character as good or evil.[8] Augustine then writes *Furtum certe punit lex tua, domine, et lex scripta in cordibus hominum, quam ne ipsa quidem delet iniquitas: quis enim fur aequo animo furem patitur*. "Thy law, o Lord, certainly punishes theft, and the law written in human hearts, which even injustice itself does not delete, for indeed what thief suffers a thief with an even mind." The prime position of *furtum* indicates "theft" as his blameworthy object. The apostrophe, *domine* "o Lord," is the focal middle term of his sentence. It identifies and invokes God as his praiseworthy object. The apostrophe also separates the two distinct subjects of his verb *punit* ("punishes") to differentiate *lex tua* ("thy law") from *lex scripta in cordibus hominum* ("the law written in human hearts"). The dependent clause, "which even injustice itself cannot delete," modifies "the law written in human hearts," not "thy law, o Lord." The relative *quam* is singular, modifying a singular antecedent. Augustine states that even "injustice itself (*ipsa iniquitas*)" cannot delete the law of the heart, which he will relate by a chiasmus to "my iniquity (*iniquitas mea*)."[9] Chiasmus was a syntactic inversion[10] that was not only aesthetic and structural but also mnemonic,[11] befitting his recollection of his individual act and a universal norm. Chiasmus was a form of parallelism that ordered syntagmatic elements in a "specular or 'mirrorlike'" arrangement.[12] Augustine will hold up his own injustice to injustice itself by inspecting his adolescent theft in the Roman customary mirror. That mirror was comedy. As the saying went, "Comedy is the imitation of life, the mirror of custom, and the image of truth."[13]

Augustine's self-blame reprises the role of a certain character, a braggart soldier (*miles gloriosus*), the comedic type that impersonated hyperbole.[14] That impersonation befit Augustine's hyperbole of proverbial thieves' honor as punishing theft. He employs it in that sentence as a personal argument from the topic of his adolescent evils. As Aristotle's *Rhetorica* allowed, "Hyperboles are for young men to use; they show vehemence of character."[15] Quintilian's *Institutio oratoria* taught hyperbole as an audacious trope, "an appropriate exaggeration of the truth." It enhanced or debased the character of the individual under praise or blame. Although hyperbole classically served various usages, its meanings were literally incredible. However, "Hyperbole is a liar, but does not lie to deceive." It conformed to a human preference for the exaggeration or diminishment of truth. Even illiterate commoners used hyperbole when the simple truth was unsatisfactory. It was a pardonable disregard of the truth, Quintilian excused, because it did not definitely assert error. Ultimately, "Hyperbole only has positive value when the thing about which we have to speak transcends the ordinary limits of nature. We are then allowed to amplify, because the real size of the thing cannot be expressed, and it is better to go too far than not to go far enough."[16]

Augustine consistently refers "thy law" to the entire Bible from Genesis to Revelation, "all the way from the beginning, in which you made heaven and earth, up to the endless kingdom with thee of thy holy city." His inclusion of the Old Testament incorporated the heart, the principal anthropological concept of the Hebrew Bible.[17] Its heart was the agent of the law, acting for or against the divine commandments.[18] Augustine confesses his discovery as a youngster of "thy law" through his punishing grief over the death of a school-and-play mate. For, his anguish over that lost love departed from God's pleasure into his wrath. As he learned, people should be loved not for themselves but only in God. "For where does he not find thy law in his punishment?" he reflects. "'And thy law is truth.'" Augustine develops that citation of Psalm 119 by detailing his torment: how after baptism his friend recoiled from him as if an enemy and threatened to cut him off if he did not stop ridiculing that sacrament; how he was stunned and disturbed by this rebuff; how at his friend's death his heart mourned grievously and he found relief only in tears.[19] This was a friend he had earlier led astray from Catholic faith into Manichaean fiction. For, Augustine believed the Manichaean claims that the New Testament was corrupted by its incorporation of Judaic law, which it rejected. However, from the teaching of Ambrose, bishop of Milan, he became aware that the Catholic faith was defensible against those charges by a spiritual, rather than literal, interpretation of the Hebrew Bible. In his conversion to the Catholic way, Augustine assimilated belief in the Judaic decalogue, which prohibited theft. "And I used to rejoice also that the ancient scriptures of the law and the prophets were now not proposed to me to be read by that eye by which I previously used to see them as incongruous." He avidly perused those questions that once seemed self-contradictory and incompatible with the testimonies of the law and the prophets, until they blended into a single pure authority. And he learned from the psalmist to exult with trembling in their harmony. Augustine discovered that whatever he read there, with the commendation of grace was said here, in perfect agreement with "thee, who art always the same."[20] Augustine resolved to live by biblical law.

"Thus, I shall consider the wonders of thy law." In his *Confessions* he recites Psalm 119 on the divine law 39 times. "For some time," he reflects, "I have been kindled to meditate on thy law and to confess to thee my learning and unlearning in it." He beseeches God to hear his praise and to allow him to "'behold the wondrous things from thy law'" in their entirety. He declares that his desire for the justice of God's kingdom arises from the divine law. "See, my God, whence may be my desire." And he cites "'The unjust have told me delights, but not thus thy law, o Lord.' Behold whence *is* my desire." In revering "thy law," of importance is that particular verse, which Augustine inserts in the mouth of Continence. She exhorts his conversion to her virtue by parading her many converts to mock and shame his hesitation. She advises him to deafen his ears to the allurements of his unclean members. "'They narrate to you delights' but not as the law of the Lord thy God." Of his conflicted heart Augustine acknowledges, "That quarrel in my heart was none other than from me myself against myself." He prays for ample time to meditate on the secrets of "thy law" and asks God not to shut the law against his knocking, so that he may serve lawfully in fraternal charity.[21] Augustine affirms in the presence of God and the Church that the legitimate and correct use of "thy law" is for charity. "For edification the law is good, if one uses it lawfully, because its end is charity from a pure heart and a good conscience and from an unfeigned faith."[22]

Augustine then decisively contrasts God's law with human custom. He confesses his ignorance before his conversion of "the true interior justice, judging not from custom (*consuetudo*) but from almighty God's rightest law, by which the manners (*mores*) of regions and days are fashioned for regions and days, since it itself exists everywhere and always." The universal and sempiternal divine law exists "everywhere and always—not here one thing or there another." Augustine strictly distinguishes the judgment of divine law, which requires obedience not disobedience, and the judgment of human custom, which elicits conformity or nonconformity. He tells that he had approached the universal justice of divine law through learning the rules of Latin metrification. For, when he chanted poetry in school he was not permitted to place the stress wherever he liked. That prohibition of choice existed because the technique for recital did not accentuate syllables helter-skelter but in a uniform fashion (*omnia simul*). Yet, despite that technique Augustine confesses his failure to infer that divine justice, which good and holy men heeded, held everything together (*simul omnia*) without any particular variations.[23] The rules of Latin metrification were, in fact, an imperfect analogy because their universality was an arbitrary system of conventional usage.[24] But Augustine believed that musical rhythm was permanent, eternal, and equal, and so allowed the contemplation of intellectual truths by a Pythagorean method.[25] His *Confessions* states that the rhythmical laws exist in the memory. "Memory contains the calculations of rhythms, and measurements, and innumerable laws that no bodily sense impresses."[26] Before his conversion these measured rhythms are nearest in Augustine's experience to "thy law." However, he complains that studious practice did not induce him to understand the analogy.[27]

In his advancement from pupil to rhetor, the reputation of Milan for a certain orderliness in education draws Augustine there. His complaint is against his former students in Carthage, who did as if lawful what was never allowed "by eternal law." Augustine judges they should have been punished legally for their misbehavior. Instead, human custom (*consuetudo*) excused them. Human custom is the butt of Augustine's criticism. Although he was personally punished as a schoolboy struggling to learn Greek, the punishment contradicted the freewheeling curiosity that learns language from speakers, rather than by compulsion from teachers. "But the flow (*flux*) of it restrains this by thy laws, o God, by thy laws from the canes of teachers up to the trials of martyrs, by thy medicinal laws to compound a healthful bitter pill, recalling us to thee from the pernicious pleasure by which we departed from thee." The flow of human custom (*mos*) away from God is "the hellish river" into which men throw their sons with tuition fees, so that they might learn rhetoric from such a master as Augustine. So there is much ado, he complains, about oratory declaimed publicly in the forum. His students were not meditating on "thy law" but purchasing from his mouth raving lies and forensic battles. They did not read about the true God because imperial law, Julian's edict, forbade Christians to teach grammar and rhetoric. Instead, Augustine's students adrift among pagan teachers read plays about pagan Jove thundering and adulterizing, so that the false thunder of their oratory could pander to humans committing adultery in imitation of that god.[28]

Thieves

It is a Roman paradigm of divine immorality, Jove's thunderous rain as semen, that Augustine and his companions imitate in the theft of pears. *Furtum certe punit lex tua, domine, et lex scripta in cordibus hominum, quam ne ipsa quidem delet iniquitas: quis enim fur aequo animo furem patitur.*[29] Theft (*furtum*) was forbidden by the biblical divine commandment "You shall not steal."[30] Theft was also a delict in Roman law, as old as the Twelve Tables, which condoned killing a thief, if caught in the act at night.[31] Schoolboys chanted that law.[32] But the details of the nocturnal episode in which Augustine steals pears to sample then throw at pigs are not literal. In a parody of the pastoral genre the setting at the neighbor's farm imitates the archetypal garden in Homer's *Odyssey* where "pear matures on pear." In classical bucolics a pear was an erotic symbol for a sexually "ripe" person, physically mature for intercourse. Augustine reveals his father's prurient notice of him at the public baths, naked and pubescent. He recounts his sexual exploits, both real and made up, to compete in obscenity with other boastful boys. However, in Roman animal husbandry there were no pigs abroad at night. At dusk they were safely confined: not even penned together but separated in colonnaded sties lest they lie on top of one another and cause the sows to abort. Augustine and his fellows would have needed to circumvent the guardian swineherd to toss pears to each separated pig. Moreover, "pigs" was classical slang for the female genitals mature enough for legitimate sexual intercourse toward pregnancy. As the linguist Varro explained in *De agri cultura*, "Our women, and especially our nurses, call that part which is in girls the mark of their sex *porcus*, as Greek women call it *choeros*, meaning thereby that it is a distinctive part mature enough for marriage."[33]

Augustine's *furtum* is not even the "touching" or "handling" (*contrectatio*) that defined the delict of theft in Roman legal texts. He employs the verb *contrectare* only for the sense of touch itself, notably God's healing touch. Nor is there in his examination of conscience about his motive for his theft the semantics of the mental state legally required for theft, *dolus*, or even *animus furandi*.[34] Augustine's single *dolus* is for his trickery as a rhetor, how he used to teach law students the loopholes to acquit guilty persons of capital crimes. For his theft of the pears he uses only the lay word *auferre*. He mentions legal "theft (*furtum*) from another house, if the occasion arises," in parallel with "committing adultery."[35] But although Augustine names the "theft" of pears, its episode belongs to his confession of adolescent fornications. He introduces the collaborative theft as a prolongation of his "sport from contagious custom" (*ludum de pestilentae more*) in the streets with his age group.[36] This is the *ludus* that was for Roman males the period of sexual initiation and experimentation between puberty and youth. Since males married only around age twenty-four, their interim recourse for sex to courtesans, slaves, or concubines was both expected and permitted. Forbidden only were adultery with another man's wife and the passive position in a homosexual act.[37] Augustine details the steamy, slimy concupiscence of his oversexed puberty bubbling in fornications. His theft climaxes his account of "gross and scummy lust" (*libido*) for shameful sexual conduct (*flagitia*).[38] His behavior was normative by Roman law and custom, and that is his criticism.

Augustine's *furtum* "theft" is literary *furtum*, theft as "rape." And that is why he is anguished and analytical about his sin. It is not the peccadillo of common interpretation, pilfering a neighbor's fruit. Latin poetry styled as *furti*, "thefts," the rapes of Jupiter/Jove in his infinite deceitful guises by metamorphosis. His "thefts" were poetized in Propertius's *Elegiae* and sweetly to Cynthia[39]; in Ovid's *Metamorphoses*, which celebrated more than fifty rapes[40]; and with human comparison in Catullus's *Carmina*, where the poet endured the thefts of his Lesbia just as Juno stemmed her anger at Jove's thousand-and-one trysts.[41] Augustine signals his own lustful metamorphosis in writing about his sixteenth year, how he was "growing to wood (*siluescere*) in dark and divers loves."[42] His verb *siluescere* derives from Calcidius's commentary on Plato's *Timaeus*, where *silva* translates *hyle*, "matter." It connotes ugly and unintelligible evil, gloomy chaos reverting to the primordial void before creation.[43] Augustine, imitating Jupiter's stealthful virility, is materializing by and from flesh into wood, so that he blends in with the pear tree of the theft.

Augustine *adulescens* identifies as a thief/rapist with Chaerea, the protagonist *adulescens* of *Eunuchus*, the most popular comedy of Terence.[44] Chaerea is sixteen, so is his girl, Pamphila; and Augustine three times dates his own theft as the deed of his sixteenth year.[45] Chaerea is upbraided by other characters for raping Pamphila, a virgin (*virginem vitiare*, *vitio virginis*).[46] Augustine blames his personal theft as vicious, asking "What in that theft did I love and what, o Lord, both corruptly (*vitiose*) and perversely have I imitated?" He twice borrows from that comedy the word *facinus*, "outrage," to blame his theft, "o my theft, and my outrage" (*furtum … facinus*). He repeats its word *flagitium*, "shameful deed." And he calls himself "shameless" (*impudens*) and a "monster" (*monstrum*) just like Terence's protagonist.[47] The female slave, Pythias, pulls no punches in calling Chaerea "a god-forsaken criminal" who "dared to do a brazen deed" (*audax facinus facere*). The courtesan, Thais, interrogates him:

Thais.	"What have you done?"
Chaerea.	"Nothing very much."
Pythias.	"Hey, nothing very much, you shameless creature (*impudens*)? Does it seem to you nothing very much to rape a citizen girl?"
Chaerea.	"I thought she was a fellow slave."
Pythias.	"A fellow slave! I can scarcely restrain myself from flying at your hair, you monster (*monstrum*)! (*to Thais*) On top of it all he comes here to mock us."[48]

Chaerea's alter-ego, the eunuch whom he has impersonated, exclaims, "Damn it! For goodness sake! I've never even heard of such an unspeakable act (*infandum facinus*), my dear." In a case of mistaken identity, the eunuch is castigated as a "villain" and a "gallows-bird."[49] As an effeminate male, a eunuch was a disgraceful figure,[50] yet Augustine regrets that as an adolescent he himself did not become a eunuch for the sake of God's kingdom.[51]

In Terence's plot Chaerea, a soldier A.W.O.L., while hurrying through Athen's streets to plan a party, is smitten at first sight of a shapely slave girl. He describes her as a luscious fruit, "real color, firm body, and plenty juicy." Chaerea orders his personal slave to procure her, in the three legal terms for improper possession of property. "Now get her delivered to me, by force or stealth or entreaty. I don't mind how, so long as I get possession of her." His slave warns him, "We're committing an outrage (*flagitium*)." But Chaerea rationalizes his trickery by Terence's ethics of "fair and right (*aequom*)" as payback to all women for their trickery. He retorts, "Is it an outrage (*flagitium*)?" By impersonating the eunuch who has been hired to attend the girl, Chaerea gains entry into the house of Pamphila's adoptive sister, the courtesan Thais, and into its private women's quarters. There he shuts the door against intruders and rapes Pamphila. He had not foreseen the deed, he claims. Initially he only sought entry to the house "to look at, listen to, and live together with the girl I wanted." Yet, his bursting joy at that prospect was sealed under oath "by Jupiter," which presaged that god's intervention as his mentor. For, Chaerea's instigation and justification for his rape was Pamphila's bedroom mural depicting Jove's rape of Danaë in a rain of gold.[52]

As Chaerea brags about his sexual exploit to a friend, "The girl sat in the room, looking up at a painting; it depicted the story of how Jupiter sent a shower of gold into Danaë's bosom. I began to look at it myself, and the fact that he had played a similar game (*luserat*) long ago made me all the more excited: a god had turned himself into human shape, made his way by stealth onto another man's roof, and come through the skylight to play a trick (*ludus*) on a woman. And what a god! The one who shakes the lofty vaults of heaven with his thunder! Was I, a mere mortal, not to do the same? I did just that—and gladly." After Pamphila bathes, then returns to her bedroom, Chaerea is ordered by the maidservants to fan her dry for her nap. After they leave, he peeks through the fan at her naked body, bolts the door, and rapes her. As he tells his friend, "Was I going to let slip the opportunity when it was offered to me, so great, so fleeting, so desired, so unexpected?" Had he not complied with Jove's example, he rationalizes, he would indeed have been the eunuch he impersonated.[53] Chaerea swaggers as the braggart soldier, the stock character of comedy who impersonated hyperbole.[54] Augustine's hyperbolic self-blame finds its match.

Rape was often a premise for comedic plots, but this rape differed in being part of the action, thus characterizing the rapist's motive. Chaerea is "a novel persona" in his lust to violate the virgin without any consideration of morals or consequences.[55] Yet, Terence's first commentator, the grammarian Aelius Donatus, mitigated Chaerea's violence by comparing Pamphila's regular bath with the ritual bath for the marriage ceremony.[56] A modern characterization admires Chaerea as "one of the most charming scapegoats in all comedy ... impulsive, passionate, tender, resourceful, manly, pious, true, a Catullus in action, scandalously indecorous, irresistibly loveable." His rape was "a mad prank."[57] In a similar opinion, "Chaerea is an engaging scamp, willy, frank, and ebullient, and it is easy to enjoy his ingenuous elation, despite the injustice." Open empathy with him is "licensed by the holiday mood of the comedy, as well as by the custom of the genre."[58] Classical literature and art frequently portrayed the women Zeus/Jupiter impregnated not as victims in pain but as consorts in pleasure toward the birth of a heroic child.[59] The Christian apologist Arnobius dismissed Danaë as a girl who could not keep her virginity; thus a theft by Jove was concocted.[60] Yet, even in poetry, some of Jove's loves were victims, frightened women who cried out and fled to escape him.[61] Danaë, the model in the comedy, was unable to flee Jove. She was imprisoned in a bronze tower by her father for his own safety, to thwart an oracle that her future son would kill him. While wondering at the historical responses of a Roman audience to theatrical rape,[62] literary criticism has also expressed revulsion at Chaerea's delinquency. It has demanded the banishment of euphemisms for his rape, "seduction" or "sleeping," which implied Pamphila's consent, for plain language: "call it rape."[63]

As the servant Pythias observes sympathetically, "On top of it all, after he'd had his fun and games with the poor girl (*ludificatust virginem*), the villain ripped her whole dress and tore her hair." In revenge Pythias wants to scratch his face with her nails, "the poisonous wretch." She reports that "the girl is crying and doesn't dare say what happened when you ask her."[64] Pamphila is a *virgo*, the stock character of comedy who is an unmarried girl of morals. Even after her rape, she is called *virgo* in deference to her moral character.[65] Yet, she is a cipher who never speaks. Her name does not even appear in the formal cast of characters for the play. As the plot devolves, however, Pamphila's identity is discovered as a freeborn Athenian citizen, not a slave. The injury to her honor must be righted by Chaerea's marriage to her, the happy ending that defines the play as comedy. Jupiter, the exemplar of the rape, is beseeched for their marital blessings.[66]

Augustine inveighs broadly against the obscene mores that such rhetoric promotes, mores he blames for his own sexual sins.[67] His *Confessions* about the theft of pears already quoted that very text of Terence's *Eunuchus* about the rape of Pamphila. As he remembers his schooling, he writes, "Thus, I truly would not have known these words 'rain' and 'golden' and 'lap' and 'deceit' and 'temple of heaven' and the other words written in this passage, unless Terence had introduced a good-for-nothing adolescent, proposing to himself Jove as an exemplar of illicit intercourse (*stuprum*), while he gawked at a certain mural, where this scene was: as the story goes, once upon a time Jove acting the fiancé sent into the lap of Danaë a golden rain to play a trick on the woman. Just look at how he arouses himself to lust, as if by a heavenly teaching: 'And what a god!' he says. 'Who shakes the temples of heaven with the loudest noise. Should I a mere mortal not do it? I truly did it and freely.'"[68] This double exemplar of Jove and Chaerea excited Augustine the schoolboy with permission to fornicate. Terence's

Eunuchus was performed at games, the Megalensian Games (*ludi*) in honor of the Magna Mater, as its production notice states.[69] Such theatrical entertainments, as stylized displays, merged in a complex cultural ritualization with play (*ludus*). Roman "play" meant school, especially rhetoric school, because it prepared boys by declamation for serious civic engagement in the forum or the court. In transition from the family to the state, adolescents were rehearsed in performance, especially by imaginative, fantastic play, which conferred authority when they were graduated into the real world. The effective role of play associated it with children verging on adults, adolescents maturing physically and socially. Play was a bodily faculty, and it embraced their pubescent sexual experiences.[70]

Augustine's theft of pears acts out the *ludus*, the customary Roman period of sexual sporting, of fooling and playing around.[71] It imitates Jove's fooling (*luserit*, *eludere*) with Danaë in Ovid's *Metamorphoses*,[72] which was imitated in turn by Chaerea's trickery (*ludificatust*) of Pamphila in Terence's *Eunuchus*. Later in *De civitate Dei* Augustine will rant against the religious castration of eunuchs and the Roman preference for the imitation of lustful Jupiter above the teachings of Plato or the values of Cato. It will cite that same comedy his *Confessions* blames. "Thus in Terence that profligate adolescent saw a certain panel painted on the wall: 'where there was this picture, as the story goes, about Jove by betrothal of Danaë once upon a time sending into her lap a golden rain.' And from this so great an authority he summons patronage of his foulness, since he boasts that in doing it he imitated a god. 'And what a god! he said, who shatters the temples of heaven with the loudest noise. Am I, a mere mortal, not to do that? I truly did that and willingly.'" Augustine's invective against theatrical displays and the worship of their gods questions: "And was it proper that your Terence should excite adolescent lewdness with the disgraceful deed (*flagitio*) of best and greatest Jove?" Augustine replies, "I conclude for the Christian: in no way, therefore, should such gods be cherished."[73] He repeats the example in a letter, displaying how much his imitation of Jupiter's rape weighs on his conscience. "And in truth, that adolescent of Terence, who looking at the mural painted on the wall, where the picture was of the adultery of the king of the gods, which seduced him to lust, inflamed by such great authority: in no way would he have either sunk in lusting, or stuck to perpetrating that shameful deed, if he had not preferred to imitate Jove, rather than Cato." Augustine excoriates Roman culture whose arts multiplied Jupiter's adulteries to compel his worship in the temple and enjoyment in the theater.[74]

Augustine's *Confessions* inserts his own theft into that pagan tradition. "What have I imitated?" he asks, reflecting on "that theft, my nocturnal outrage in my sixteenth year." His imitation was of Chaerea raping Pamphila in imitation of Jove raping Danaë. Jove's mythological downpour of gold repeats in Augustine's imaginative throwing of the golden pears. The shower of semen flows down from Jove fornicating into Augustine's "whirlpool of shameful deeds," where he spins out of control until he is sucked into its vanishing point. This is the "flow (*fluctus*) of my age" into "the flood of human customs,"[75] he acknowledges. Among Roman moralists, Seneca conceived sensual pleasure as a fluid lapping at the body and seeping into its pores to invade and seduce the mind.[76] Among the meanings of the wood (*silva*), into which Augustine at age sixteen has metamorphosed, was the flowing, fluctuating river (*fluxus*).[77] Augustine confesses to God, "But wretched I boiled over; following the rush of my flowing (*fluxus*), I left thee behind; and I departed from all your lawful prescriptions but I did not escape your whips, for what mortal can?" Into that "hellish flood" of custom fathers tossed their sons with fees—Jove's lucre—for their education in rhetoric.[78] The grammarian Donatus commented that the courtesan Thais's household mural of Danaë's rape was "against the customary shame of all persons," and he associated its golden rain with golden coins, for her avarice in selling sex.[79] It is during an impecunious period, when Augustine's father strives to earn or borrow the tuition fees for his son, that Augustine on vacation from school samples the fruits of his rhetorical education. "Thus," he reflects to God, "does the soul fornicate when it goes away from thee and seeks without thee those things that it does not find pure and liquid except when it returns to thee." It is with "a heavy rainstorm of tears" (*ingentem imbrem lacrimarum*), reversing Jupiter's golden rain of semen (*imbrem aureum*), that repentant Augustine streams his soul back to God.[80] He converts from a boundless lack of self-control to Continence, a stern Roman virtue.[81]

In recollection of his injustice in the theft of pears, Augustine's address to God, *domine*, redeems the oath of Chaerea, *pro Iuppiter*, gleeful at the prospect of having Pamphila.[82] Augustine rehearses the commonplace criticism of Roman moralists that the theater promoted sexual immorality.[83] He criticizes Terence's comedy for exciting lust by its language and by its mural.[84] Augustine believed the medical fallacy that the mental fantasy retained the images it viewed, so realistically and efficaciously so that a fetus bore traces of the object on which its mother passionately gazed while conceiving.[85] At issue for his confession of the theft is pagan exemplarity, promoted by Roman custom in defiance of "thy law." He would have agreed with the modern critic who characterized Chaerea the rapist as "pious." That is precisely Augustine's condemnation of Roman rhetoric that it glorifies as godly "playful" sexual violation. It lauds Jupiter, the pagan ideal of male potency in his prowess to trick humans for gratification of his insatiable lust.[86] Augustine and his schoolmates know from declamatory exercises that rape is a statutory crime subject to capital punishment.[87] But in the play Chaerea comically excuses his rape because he thinks Pamphila a slave,[88] without legal status or social honor. Augustine's thrust of pears at pigs on the neighbor's farm plausibly symbolizes his sexual "fooling" particularly with slave girls. As a bishop, he will later preach against the Roman civil law that classified slaves as the property of their masters. He will declare that the sexual acts of masters with their slaves are sinful under divine law, "thy law."[89]

Outlaws

Augustine develops the injustice of his theft by a comparison that broadens from literary theft as rape to legal theft as outlawry. *Furtum certe punit lex tua, domine, et lex scripta in cordibus hominum, quam ne ipsa quidem delet iniquitas.* "Thy law, o Lord, certainly punishes theft, and the law written in human hearts, which even injustice itself does not delete." He proves his claim about "the law written in human hearts" thus: *quis enim fur aequo animo furem patitur?* "for indeed what thief suffers a thief with an even mind." His phrase *lex scripta in cordibus hominum* "the law written in human hearts" might seem a biblical allusion. In the Epistle to the Romans "the nations (*gentes*), who did not have the law, naturally (*naturaliter*) did those things that are of the law. In such a manner, not having the law, they themselves are a law to themselves, which shows the work of the law written in their hearts (*opus legis scriptum in cordibus suis*)."[90] Augustine's *Confessions* cites other verses of that epistle, dramatically so for his conversion under a fig tree in a Milanese garden.[91] But it does not cite Paul's verse about "the work of the law written in their hearts (*opus lex scriptum in cordibus suis*)," meaning the Gentiles, who lack God's law. Augustine writes of "the law"—not its work—"written in human hearts" (*lex scripta in cordibus hominum*) universally. His *Expositio* of some propositions of that epistle does not cite "the law written in their hearts" either. It does cite the continuation of the verse, "with their conscience as witness," but with reference to another biblical verse Augustine identifies "thy law."[92] However, Augustine's response to Faustus, a Manichaean bishop who figures in *Confessions*, cites Romans 2:14 for Jesus's fulfillment of the law and the prophets.[93] It adduces the verse for the second of three types of law: "Jewish, which Paul calls of sin and death; Gentile, which Paul calls natural; and the truth, which Paul calls the spirit of life in Christ Jesus."[94] However, Augustine's "law written in human hearts" differs substantially.

Augustine's *Confessions* does not define it as natural law. Its only occurrences of *naturaliter* "naturally, by nature" do not repeat Paul's word. Augustine's diction is not about the observance of any law. He confesses, rather, his Manichaean belief that he himself was like God, "what you are naturally,"[95] then his Christian learning that the Father and the Son are the same "naturally."[96] Those uses of *naturaliter* "naturally" correspond to Aristotle's fundamental logical category, "substance," the predication of what something is.[97] Augustine's *Confessions* reports that, at scarcely age twenty, Aristotle's *Categories* came into his possession. He boasts that he "read and understood it by himself," specifically without assistance from his rhetoric master or other learned expositors. "And it seemed to me to be speaking clearly about substances, such as what is man," and the other nine categories predicated of man. Augustine confesses that he subjected God, as if he were bodily like man, also to Aristotle's categories.[98] His Manichaean belief about what God is "naturally" concerns Aristotle's distinction of a secondary substance, because of a shared materiality of bodies, divine and human, God's and Augustine's. His Christian belief about God "naturally" concerns a primary substance because Augustine learned that Father and Son are the same unity. Augustine's usage of *naturaliter* there is logical, not legal.

Augustine's amplification of his sentence explains his distinct meaning of "the law written in human hearts" that punishes theft. *enim fur aequo animo furem patitur?* "for indeed what thief suffers a thief with an even mind?"[99] Augustine's law in human hearts is social affinity, proved by proverbial thieves' honor, "for indeed what thief suffers a thief with an even mind." Thieves' honor was no natural law that observed God's law in its absence. It was under Roman law an outlawry—outside not only divine law but also outside human law. It was neither positive law nor natural law but tradition or custom that devolved to social practice. It was not about legal obedience or disobedience but about social conformity or nonconformity. Roman piety respected *mos maiorum*, the ancestral tradition, as the foundation of the state. Although its norms were somewhat internalized socially, they were arbitrary, not derived from a universal law or principle.[100] Augustine regrets a loss of the ancestral values in their best moral practice. He contrasts them with "the customs of my parents," which engendered "our custom," his adolescent group at play.[101]

Augustine's phrase about proverbial thieves not suffering a thief *aequo animo*[102] recalls and reverses the comedic Chaerea's perverse rationalization of his rape from Terence's ethics of "the right and good (*aequom*)." Chaerea justifies his intended "outrage" (*flagitium*) of deceitfully invading the courtesan's household by declaiming the universal chicanery of women.[103] His motive is tit for tat. This egalitarianism parodies the golden rule, "Do unto others as you would have others do unto you." Chaerea cunningly intends revenge: to "get even" (*aequare*) with women in general, not to do one girl, Pamphila, justice. Augustine's proverb about thieves' honor amplifies that "the law written in human hearts" acts as a law unto oneself in an anarchic sense. He confesses that his lust exceeded the civilizing boundaries of friendship and marriage. It was impelled by the perverse "social necessity" of Roman custom—not law—which allowed and expected adolescent sexual play (*ludus*). Augustine emphasizes that he would not have committed the assault alone. He did it for the applause of his peers. "But when it is said, 'Let's go, let's do it,' we are ashamed not to be shameless." As a comedic performance, his gang assault on pigs with pears was "a laugh that tickled the heart" because it deceived those who thought the boys would not do such things. It was a funny trick, he writes, much better enjoyed in company than by laughing alone. "Alone I would not have committed the theft because I did not enjoy what I stole but the fact of stealing. I would not have enjoyed doing it alone nor would I have done so." Their nefarious deed was "a mysterious seduction of the mind, an unfriendly friendship, craving to inflict harm from play (*ex ludo*) and by a joke." Augustine's ludic pleasure in the theft of pears was "in the outrage (*facinore*) itself, which the company of sinners made together."[104]

Augustine's amplification of his theft resonates with the proverb "thief knows thief, and wolf wolf." On friendship Aristotle cited it in *Ethica Eudemia* among proverbs that associated like with like.[105] As Erasmus explained it in his great Renaissance collection of classical *Adagia*, "Some mutual affection is commonly found between those who suffer from similar failings, and specially among thieves; and wolves like robbers hunt in packs. We can see affection developing immediately between certain people because their natures are in some secret way akin." The fellow feeling among thieves that Augustine references belongs to a proverbial complex. As Erasmus elaborated, "Similarity is the mother of goodwill, and links people together by habit and way of life. So we see young people liking to come together … ruffians with ruffians." Such intimacy originates in an equality of age so that "adolescent and adolescent" naturally enjoy one another. This harmony echoes the adage that heads Erasmus's collection, "Between friends all is common." Erasmus cites from varied sources including the playwright Terence toward a Christian communism of shared goods.[106] Aristotle's friendship was egalitarian in identifying a common humanity from a fundamental likeness among persons. Friendship was a universal phenomenon, extending even between bad persons, although their friendship was not based on a virtuous but a practical similarity. Thieves maintained a partnership in crime toward a useful, although unlawful, end.[107]

Aristotle on proverbial thieves indicated that the physicists, meaning the Stoics, organized all of nature on this principle of sympathy.[108] Augustine borrows the hyperbole about honor even among thieves from Cicero's *De officiis*, an ethical treatise on the imitation of cosmic harmony by human cooperation. Cicero's text, addressed to his son, Marcus, away at school, represented the Roman moral instruction that Augustine's father, Patricius, culpably denies him as a schoolboy. Discoursing on duties Cicero thought that a reasoned, educated life desired the assistance of familiars, whom it attracted by personal justice. This cooperation was a required virtue even for a country dweller (like Augustine at the family farm). Cicero declared justice necessary also for buyers and sellers, hirers and leasers—for all commercial transactors. "Its effect is so great that not even those who win their bread from evil-doing and crime are able to live without any particle of justice. For if anyone steals or snatches something from one of his fellows in banditry, he leaves no place for himself even within the gang of bandits. And if the one called the pirate chief does not share the booty fairly, he will be killed or abandoned by his comrades. Indeed they say that there are even laws (*leges*) among bandits, which they obey and respect." Cicero adduced two bandits who acquired great influence through their fair distribution of the loot. He summarized, "Justice has such great effect that it strengthens and increases the resources even of bandits. How great an effect, then, do we think it will have among laws and lawcourts and in a well ordered political community?"[109]

The passage has been commented an a fortiori argument, a logical inference by stronger reason from a lesser fact to a greater probability.[110] Cicero's example was only superficially logical, however. Although in modern usage a commonplace is a trite saying, in Roman rhetoric a commonplace was a technical argument. An amplification, it exposed evildoers in black-and-white judgment. Epideictic rhetoric praised or blamed an individual, such as God or Augustine. A commonplace, however, blamed a type: the thief, the adulterer, the tyrant, and others.[111] Augustine acknowledges thieves' honor as a truism by posing his amplification on his theft as a rhetorical question, "for indeed what thief suffers a thief with an even mind." The conjunction that introduces his amplification, *enim*, is a demonstrative particle that was classically employed to corroborate its preceding clause. Its confirmation meant "truly, certainly, to be sure, indeed, in fact." However, Cicero's commonplace about justice among bandits and pirates, on which Augustine depends, was a hyperbole. With it Augustine amplifies his own hyperbole about the irradicability of the law in human hearts even by injustice itself.

Although Cicero is hailed as a popularizer of Stoic natural law, Augustine never would have accepted his norm, "The true and chief law proper for commanding and forbidding is the right reason of supreme Jove."[112] Cicero addressed its opposition by outlaws—bandits on land and pirates at sea. Although bandits could be protected, even heroized, by the peasant fellows from whom they emerged,[113] Cicero hardly admired bandits as models of just citizens. Indeed, he indignantly upbraided his contemporary citizens for their immorality in allowing pirates the license to operate without reprisal, while they subjected allies to taxation.[114] As governor of Cilicia,[115] Cicero knew of the nefarious Cilician piracy, which surfaced to menace commercial shipping and to auction captive citizens for slavery. Its only record of religiosity was the "strange sacrifices" and "secret rites" conducted by one pirate chief to the god Mithras.[116] Cicero's writings were the principal witness to piracy at its Roman apex, and his bandits figured not as doers of any inherent natural law but as rankly immoral outlaws. He paraded pirates in chains and he reveled in the public cheer at their gory executions. His prosecution against Verres quotably censured pirates as the consummate evildoers, "the common enemy of all peoples."[117] Augustine reckoned the Verrine orations Cicero's noblest.[118] His *De officiis*, Augustine's source for honor among thieves, further stated his judgment that a pact with a pirate as the price of one's life, even if given on oath, was not binding. "For a pirate is not counted as an enemy proper, but is the common foe of all."[119] This judgment of Cicero's was the clearest statement of the attitude of the Roman elite toward pirates as undeserving of the respect owed even to enemies of state under the conventions of war.[120]

As outcasts, Cilician pirates established their defiant counter-society on a justice that was egalitarian. However, their rule of equality developed practically, not theoretically or legally. It reacted to historical maritime experience, to the hierarchy of merchant ships with its ranked privileges and abuses, and to the scarcity of supplies aboard with its contingent deprivation of necessities to the crew. Within their bands pirates expected fair-and-square behavior among the members. Toward external bands they created relationships in a network bonded on occupational cohesion. Crew members equally elected from within the ranks their chief, who would serve not only as a ruthless plunderer of outsiders but also as a decent administrator for insiders. The organization of piracy was written in loyally sworn pacts, which governed and disciplined work and war. The prized loot was portioned by the captain to the crew according to time-honored maritime shares, with some booty reserved in a common fund for the disabled and dependent members of the piratical community.[121] This band of brothers behaved justly among its own kind not by any natural law written in hearts, however. Their ethos of fellow feeling related to Aristotle's proverb "a thief knows a thief" and related proverbs expressing a cultural value of similarity, or like likes like.[122] Toward their prey and their hunters, of course, pirates exhibited no justice whatsoever. They were merciless, violent and vengeful, selling captives into slavery or shoving them overboard to drown.[123]

Banditry, of which piracy was the maritime version, was a common and harsh intrusion on daily Roman social life. It affected legal acts, as tombstones recorded death by brigands and pirates; or it was regarded among natural disasters, for which there was no legal redress. The significant fact about banditry for understanding Augustine's argument from honor among thieves is that banditry was not unlawful but *lawless*. Bandits were classified in Roman law as "common enemies of the nation" in distinction to "an enemy of a just man." Bandits were not recognized by the Roman state. As stateless, bandits occupied a no-man's land between persons under civil and criminal law, and enemies of the state. Bandits were literally "out-laws." The term *latrocinium* encompassed any "violent opposition to established authority short of war." Bandits were strictly distinguished from criminals such as common thieves. Laws in the codes and in charters empowered towns to hunt, attack, capture, and hand over bandits to the provincial governor's court. Private citizens were authorized in self-defense to employ force to injure, even kill, bandits. Avengers were exempt from legal penalties by a grant of public vengeance. Not only in the law but also in upper-class judgment bandits were grouped among the other outcasts, the slaves and the insane.[124]

The behavior of pirates differed from the maritime counter-culture of ordinary sailors—deprived of liberty and autonomy, oppressed by hard and excessive labor—who also retaliated against landed society and exemplified their civil disobedience by theft. Yet the antagonism of seamen stayed within the boundaries of social order, whereas pirates challenged it by disrupting the shipment of foodstuffs and thus threatening public survival. Since piratical assaults were indiscriminate, even on the lives and livelihoods of ordinary seamen, pirates were indiscriminately despised. Because of this universal hostility, pirates as outlaws were subjected to severer punishments than were common thieves.[125] Whether pirates at sea or bandits on land, outlaws were denied legal rights, even the rights of criminal defendants. Bandits were tried outside the normal procedure, at the will of the magistrate or governor interpreting traditional norms. Bandits were disallowed any courtesy, and they could be interrogated, regularly under torture, before sentencing. Penalties for bandits differed from those for criminals in degree and in deterrence. Judgment, as both retribution and as terrorism, sentenced them to the severest punishments: exposure to beasts, burning alive, and crucifixion. The corpses of the worst offenders were impaled on stakes at the site of their misdeeds. Such publicity and such violence exceeded the punishments of criminal law, a statement that banditry was subject to the power of the state, rather than to civil action.[126]

These realities of banditry/piracy grounded Augustine's hyperbole about the law written in human hearts as thieves' honor. His *De civitate Dei*, which historians of banditry cite as authoritative, would state Augustine's knowledge of thieves' honor in sharing bounty. "With justice removed," he argued, "what are kingdoms but huge banditries? Because what are banditries but tiny kingdoms? The legal power of the men is governed by the chief's command and bound by a social pact, and the loot is divided by an agreeable law." Augustine judged the pirate chief who marauded in a flimsy ship no different from Alexander the Great whose mighty armies warred on the entire world.[127] Augustine's introduction to this argument, like that in his *Confessions*, was his personal attendance at the theater, where the fictional crimes of the pagan gods were performed "to be imitated as if by divine authority."[128]

Augustine's *Confessions* names his sexual play at his neighbor's farm a "theft," in imitation of Jupiter's euphemistic thefts. But he associates his injustice more forcefully with banditry. He does so to accuse his "theft" as worse than unlawful: as lawless. And he does so also to expose his cheap motive for the theft. Piracy was not only a resort of the economically desperate in need of sustenance but also a lure for thrill seekers after "excitement and risk."[129] What provoked the Roman public most to anger and hatred toward the Cilician pirates whom Cicero condemned was their insolent relish in their lifestyle: outlawry flaunted at sea by their gilded sails and silvered oars, reveled in port in drinking bouts, song fests, and lusty wenching. As Plutarch moralized, they were contemptible most of all not for their evil deeds but because they "rioted in their iniquity and plumed themselves upon it."[130] Augustine's pride in the theft of the pears he discerns as transgressing illegality to outlawry. About those pears tossed to pigs, he writes, "I plucked them so I might steal." He enjoyed breaking the law "for no other reason than that it was against the law." As he blames himself for loving the deed, "I loved to perish (*perire*)," echoing the fall of the pears (*piri*).[131] Augustine the thief falls prey to himself, for *perii!* was the standard cry in comedy of the victim of theft. It meant literally "I'm ruined!" but often the expletives "damn it!" "hell!"[132]

Augustine's appeal to proverbial honor among thieves amplifies his fundamental hyperbole, "not even injustice itself can delete the law written in the hearts of men, for even a thief does not suffer a thief with an even mind." This concept is not a natural law. Collective theft as banditry on land and piracy at sea devolved to the primitive Mediterranean practice of plundering neighbors for the economic production of personal gain and communal sustenance. The historian Thucydides reported it as an ancient employment that was not yet disgraceful, while Aristotle classified brigands matter-of-factly among hunters, farmers, and fishers, much as he defined war politically as a means of acquisition different from trade. Such theft was communally sanctioned, provided that it was practiced on a foreign or inimical tribe. Theft was communally forbidden only within one's own tribe. It was to this collective mentality, before various means of acquisition were distinguished,[133] that proverbial honor among thieves pertained.[134] It was not evidence of a universal natural law, only of a historical social practice. Once theft was legally established as a crime, bands of thieves continued to observe their self-imposed law against theft among themselves. But, banded together, they robbed in order to break the statutory law against theft toward outsiders. And so they became the ultimate outsiders, outlaws. This was a conspiracy, a pact for the survival and success of a predatory group. It was not rational behavior, not a reasonable application of an inherent natural law. It was a matter of social affinity for the preservation and cohesion of a band that existed precisely by and for overturning the external social order.

Piracy was ordinary reality occasioned by the geography of the Mediterranean region, which drove peoples from inhospitable lands to earn a livelihood on the seas. It was tacitly tolerated by governmental complicity, for its delivery of the slaves necessary to the Roman economy and for political protection against the resurgence of foreign states.[135] Augustine, as the son of the decurion Patricius, collector of the annual taxation of the local crop (*annona*), knew of the piracy that traditionally threatened grain shipments to Rome from his native North Africa, cereal bowl of the empire.[136] After relative calm at sea during the Roman peace, during the late-dissolving Roman Empire piracy was again churning the waters and endangering that North African coastline. Vandals would in 429 C.E., just before Augustine's death, capture Carthage and conscript its *annona* fleet for piratical raids.[137] Augustine's writings also bristle with dangers to travelers on land from brigands. A special scourge of provincial North Africa were the sectarian circumcellions, who besieged rural farmhouses and bedeviled his episcopacy.[138] In *Confessions* he classifies banditry against a traveler among the "fundamental principles of injustice (*iniquitas*)," and he contrasts the perils of ambush by bandits to the security of walking in the Lord's ways.[139]

Banditry was broadly imported into Roman culture. From boyhood Augustine likely played *Latrunculi*, "Little Bandits." It was the most popular board game, an early chess in which the playing pieces were strategically captured.[140] Banditry was also an academic commonplace, as textbooks described the capture and torture of bandits and their march to trial through the municipal forum.[141] Declaimed in the rhetoric schools was the commonplace question of a son's obligation to ransom his parents from pirates. The elder Seneca's *Controversiae* proposed the thesis whether a son should leave his blind mother to ransom his father, while Aristotle's *Ethica Nicomachea* wondered whether a ransomed captive should ransom his ransomer or his father. Piracy also occasioned declamations about the validity of oaths to pirates sworn under duress, such as Cicero's *De officiis* where pirates were definitionally "the common foes of all mankind." Such declamations, invented from a philosophical thesis, were the standard exercises to try the skills of adolescents, the culmination of a Roman boy's education.[142] Augustine studied, practiced, and taught them.

Piracy was a topic that also moved the plots of Roman comedy with the practice of kidnapping characters for lucrative ransom or enslavement.[143] Pamphila, the violated virgin of Terence's *Eunuchus*, arrives on the scene for reunion with her supposed sister, the courtesan Thais. Years have elapsed since pirates kidnapped Pamphila and sold her to a merchant, who then donated her to Thais's mother.[144] Had Pamphila never been kidnapped, there would have been no separation, thus no premise for reunion. That occasion locates her in the street for Chaerea to notice and desire her, the occasion that motivates Thais to find her lost citizen family for monetary gain. No piracy, no plot. Augustine's reprisal in *Confessions* of Chaerea's theft of Pamphila's virginity is metaphorically piratical, for in their violent greed pirates transgressed social norms to embody lust by wenching. A controversy practiced in the rhetoric schools presented the virgin sold into slavery by pirates who menaced her virtue.[145] Pirates were libertines, like adolescent Augustine at sexual play.

Beyond the historical Roman refusal of legal recognition to banditry, there is an anthropological explanation for Augustine's proverbial honor among thieves. Outlaws operate not by law—neither for nor against it—but by a code of honor and shame. Honor is personal valuation as socially validated. Honor as the basis of precedence fundamentally conflicts with legality because seeking legal redress jeopardizes honor by a public display of vulnerability. The behavior of a school gang or street-corner society, like Augustine's adolescent group, makes "'a law unto itself.'" That happens not because the group is *above* the law but because it is *outside* it and because the concept of honor = virtue has no claim upon its aspirations. Honor answers to peers individually and collectively, not to lawmakers, divine or human. It does not depend on, or even concern, virtuous lawful behavior. Honor is anthropologically a social, not moral, concept. It determines precedence.[146]

Comedy, the Romans said, is "the mirror of custom."[147] Augustine upholds it prudently to discern the distinction between human custom and divine law. He writes with hyperbolic personal blame, *Furtum certe punit lex tua, domine, et lex scripta in cordibus hominum, quam ne ipsa quidem delet iniquitas: quis enim fur aequo animo furtem patitur?*[148] Augustine does not lecture about moral performance or avoidance but confesses the punishment of his theft. As he expounds, punishment taught him from boyhood "thy law," as Scripture revealed. For his sin, he acknowledges, God killed him. Yet, in the absence of divine law, even a human code punished theft. Augustine's fellows would have ostracized him for not participating in the theft or retaliated against him for tattling on them. It is an anthropological finding that groups cohere and flourish by punishing deviants. On the empirical evidence, this is not the choice of rationality.[149] Such behavior devolves to social affinity. At stake for adolescent Augustine is not his integrity under any law but his reputation among his peers, under a code of honor and shame. Augustine amplifies his hyperbole about the irradicability of the law written in hearts, even by injustice itself, with the hyperbole of proverbial honor among thieves. That pact was a conspiracy to outlawry, not an observance of natural law.

Augustine interpreted Adam's original sin as the pride that dared to imitate God.[150] His own theft of pears imitated another god, Jove, and that became, in Augustine's conscience, no small sin.

Notes

1. Augustine, *Confessionum libri tredecim* 2.2.9, ed. Lucas Verheijen (Turnholt: Brepols, 1981).
2. E.g., William Mann, "Inner-Life Ethics," in *The Augustinian Tradition*, ed. Gareth Matthews (Berkeley: University of California Press, 1998), pp. 157–60.
3. E.g., Gerald Bonner, "Augustine's Doctrine of Man: Image of God and Sinner," *Augustinianum* 24 (1984): 496; John Freccero, "Autobiography and Narrative," in *Reconstructing Individualism: Autonomy, Individuality, and the Self in Western Thought*, ed. Thomas C. Heller et al. (Stanford, Calif.: Stanford University Press, 1986), p. 23; Hugues Derycke, "Le vol des poires, parabolé du péché originel," *Bulletin de littérature ecclésiastique* 88 (1987): 337–48; William J. O'Brien, "The Liturgical Form of Augustine's Conversion Narrative and Its Theological Significance," *Augustinian Studies* 9 (1978): 57–58; Leo C. Ferrari, "The Arboreal Polarization in Augustine's *Confessions*," *Revue des études augustinennes* 25 (1979): 35–46; idem, "The Pear-Theft in Augustine's *Confessions*," *Revue des études augustiniennes* 16 (1970): 233–42; Robert J. O'Connell, *St. Augustine's "Confessions": The Odyssey of Soul* (Cambridge, Mass.: Harvard University Press, Belknap Press, 1969), pp. 47–50; Kenneth Burke, *The Rhetoric of Religion: Studies in Logology* (Boston: Beacon, 1961), pp. 93–101.
4. Gen. 2:18, 21–23; 3:7.

5. *Confessiones* 2.3.8–2.10.18.
6. Ibid. 2.4.9.
7. Ibid. 2.2.1.
8. Theodore C. Burgess, *Epideictic Literature* (Chicago: University of Chicago Press, 1902), pp. 115–16.
9. *Confessiones* 2.2.3–2.2.4.
10. John W. Welch, "Chiasmus in Ancient Greek and Latin Literatures," in *Chiasmus in Antiquity: Structures, Analyses, Exegesis*, ed. idem (Hildesheim: Gerstenberg, 1981), pp. 250–68. Chiasmus was also a prominent feature of the Hebrew Scriptures, whose law Augustine considered, and it appeared in the New Testament. See Nils W. Lund, "The Presence of Chiasmus in the Old Testament," *American Journal of Semitic Languages and Literatures* 46 (1930): 104–26; idem, *Chiasmus in the New Testament* (Chapel Hill: University of North Carolina Press, 1942), which cites as a rare example of extended chiasmus beyond Scripture Augustine, *Enarrationes in psalmos* 21, p. 33.
11. Ian H. Thomson, *Chiasmus in the Pauline Letters* (Sheffield: Sheffield Academic Press, 1995), pp. 34–35; Welch, "Introduction" to *Chiasmus in Antiquity*, p. 12.
12. See José Antonio Mayoral, "Chiasmus," in *Encyclopedia of Rhetoric*, ed. Thomas O. Sloane (Oxford: Oxford University Press, 2001), p. 89.
13. Cicero ap. Donatus, "De comedia" 5.5 in *Commentum Terenti*.
14. See, without reference to Augustine, Heinrich F. Plett, "Hyperbole," in *Encyclopedia of Rhetoric*, p. 364.
15. Aristotle, *Rhetorica* 3.11.15–16 1413a; *The Complete Works of Aristotle, The Revised Oxford Translation*, trans. Julian Barnes, 2 vols. (Princeton: Princeton University Press, 1984), 2:2255.
16. Quintilian, *Institutio oratoria* 8.6.67; *The Orator's Education*, trans. Donald Russell, 4 vols. (Cambridge, Mass.: Harvard University Press, 2001), 3:465, 469. For hyperbole in Roman schools, see Stanley F. Bonner, *Education in Ancient Rome: From the Elder Cato to the Younger Pliny* (Berkeley and Los Angeles: University of California Press, 1977), p. 165.
17. H. W. Wolff, *The Anthropology of the Old Testament*, trans. Margaret Kohl (Philadelphia: Fortress, 1974), p. 40, which counts 858 occurrences.
18. Marjorie O'Rourke, "'In the Heart of the Sea': Fathoming the Exodus," *Journal of Near Eastern Studies* 63 (2004): 17–27; idem, "The Law of the Heart: The Death of a Fool (1 Samuel 25)," *Journal of Biblical Literature* 120 (2001): 401–27; idem, "Broken Hearts: The Violation of Biblical Law," *Journal of the American Academy of Religion* 73 (2005): 731–57.
19. *Confessiones* 11.2.3, 4.4.7–9.

20. Ibid. 5.11.21, 5.14.24, 6.4.6, 7.21.27.
21. Ibid. 10.43.70, 11.2.2, 11.2.3, 11.2.4, 8.11.27, 11.2.4. For Augustine's text of Ps. 119, which differs from the Masoretic Text, the Vulgate version, and the Roman Psalter, see Alban Dold and A. Allgeier, *Der Palimpsest Psalter im Codex Sangallensis 912: Eine altlateinische Übersetzung des frühen 6. Jahrhundert aus der einstigen Kloster-Bibliothek von Bobbio* (Beuron: Erzabtei, 1933), pp. 30–52, with v. 85 at pp. 86, 47.
22. *Confessiones* 12.18.27, 12.25.35; 12.30.41; Deut. 6:5, Lev. 19:18, Matt. 22:35–38; *Confessiones* 12.18.27, 13.24.37.
23. Ibid.
24. See D. S. Raven, *Latin Metre: An Introduction* (London: Faber and Faber, 1965).
25. Augustine, *De musica* 6. Marjorie O'Rourke Boyle, "Augustine's Heartbeat: From Time to Eternity," *Viator: Medieval and Renaissance Studies* 38 (2007): 19–43.
26. *Confessiones* 10.12.19.
27. Ibid. 1.18.29.
28. Ibid. 5.13.23, 5.8.14, 1.16.26, 9.2.2, 8.5.10, 1.16.25. For thunderbolts of eloquence, see Quintilian, *Institutio oratoria* 8.6.4. For payments, see Bonner, *Education*, pp. 146–62.
29. *Confessiones* 2.2.4.
30. Ex. 20:15 Vulg; cf. Deut. 5:19.
31. H. F. Jolowicz, ed., *Digest XLVII.2: "De furtis"* (Cambridge: Cambridge University Press, 1940), pp. lxviii–lxxv, 1. Augustine knew this law from declamatory practice, for which see S. F. Bonner, *Roman Declamation in the Late Republic and Early Empire* (Liverpool: Liverpool University Press, 1969), p. 105.
32. Bonner, *Education*, p. 166.
33. Marjorie O'Rourke Boyle, *Divine Domesticity: Augustine of Thagaste to Teresa of Avila* (Leiden: E. J. Brill, 1997), pp. 11–12. See also Cicero, *De officiis* 1.35.129.
34. *Confessiones* 7.20.26, 10.8.13, 10.10.17, 10.12.19. For the legal terms, see Jolowicz, ed., *Digest XLVII.2*, pp. xvii, xix, lv–lxi, and theft of fruit, pp. lviii, 96.
35. *Confessiones* 4.2.2. For chicanery, see Cicero, *De officiis* 1.10.33. *Confessiones* 2.4.9, 8.10.24.
36. *Confessiones* 2.4.9.
37. Boyle, *Divine Domesticity*, pp. 9–11.
38. *Confessiones* 2.2.2, 2.3.7, 2.4.9, 2.6.12, 2.7.15, 2.8.16, 2.9.27. For scum on boiled liquids, see Aristotle, *De generatione animalium* 2.6 743b.
39. Propertius, *Elegiae* 2.2, 2.13; To Cynthia 2.30.25.

40. Ovid, *Metamorphoses* 4.611, 6.113, 11.117. See Amy Richlin, "Reading Ovid's Rapes," in *Pornography and Representation in Greece and Rome*, ed. idem (New York: Oxford University Press, 1992), pp. 157, 161.
41. Catullus, *Carmina* 68b lines 136, 140.
42. *Confessiones* 2.1.1.
43. Without reference to Augustine, see J. Reginald O'Donnell, "The Meaning of *silva* in the Commentary on the *Timaeus* by Chalcidius," *Mediaeval Studies* 7 (1945): 12–13, 8, 19, 6–7, 11–12.
44. Suetonius, *Vita* 1.
45. Terence, *Eunuchus* lines 318, 693, in *Terence*, ed. and trans. John Barsby, 2 vols. (Cambridge, Mass.: Harvard University Press, 2001), vol. 1. Augustine, *Confessiones* 2.1.1, 2.2.4, 2.3.6, 2.6.12.
46. Terence, *Eunuchus* lines 654, 704, 858, 953.
47. *Confessiones* 1.2.2, 2.3.7, cf. Sodomites 3.8.15, 2.6.14, 2.5.11, 2.6.12, cf. 2.8.16, 2.9.17, 2..6.14. Terence, *Eunuchus* lines 382, 1013, 1022, 856, 860.
48. Terence, *Eunuchus* lines 857–860; trans., p. 415.
49. Ibid. lines 654, 704, 722, 858, 953, 643–44, 664, 668, 670; trans., pp. 385, 389.
50. Cynthia S. Dessen, "The Figure of the Eunuch in Terence's *Eunuchus*," *Helios* 22 (1995): 123–39, 125; John Whitehouse, "The Rapist's Disguise in Menander's *Eunuchus*," in *Intertextualität in der griesch-römischen Komödie*, ed. Niall W. Slater and Bernhard Zimmermann (Stuttgart: M & P Verlag für Wissenschaft und Forschung, 1993), pp. 122–32.
51. *Confessiones* 2.2.3.
52. Terence, *Eunuchus* line 319. Cf. *Confessiones* for pears as "pretty" 2.6.12 but "neither in shape nor in savor alluring," 2.4.4. *Eunuchus* lines 319–20, 382–87; trans., p. 357; Barsby, ed., pp. 146, 156; line 574, trans., p. 377; line 550, cf. the anti-oath line 709.
53. Terence, *Eunuchus* lines 583–606, 289–90.
54. See note 14.
55. David Konstan, "Love in Terence's *Eunuch*: The Origins of Erotic Subjectivity," *American Journal of Philology* 107 (1986): 387; Karen F. Pierce, "The Portrayal of Rape in New Comedy," in *Rape in Antiquity: Sexual Violence in the Greek and Roman Worlds*, ed. Susan Deacy and Karen F. Pierce (London: Duckworth with the Classical Press of Wales, 1997), pp. 163–84. Terence, *Eunuchus* lines 175–76.
56. Aelius Donatus, *Commentum Terenti ad Eunuchus* line 592. Katrina Philippides, "Terence's *Eunuchus*: Elements of the Marriage Ritual in the Rape Scene," *Mnemosyne: A Journal of Classical Studies* 48 (1995): 272–84.

57. Edward K. Rand, "The Art of Terence's *Eunuchus*," *Transactions of the American Philological Association* 63 (1932): 58, 62.
58. Konstan, "Love in Terence's *Eunuch*," p. 387.
59. Mary R. Lefkowitz, "Seduction and Rape in Greek Myth," in *Consent and Coercion to Sex and Marriage in Ancient and Medieval Societies*, ed. Angeliki E. Laiou-Thomadakis (Washington, D.C.: Dumbarton Oaks Research Library and Collection, 1993), pp. 17–37; Frances van Keuren, "Female Sexuality and Danaë and the Golden Rain," *American Journal of Archaeology* 101 (1997): 369.
60. Arnobius, *Adversus nations* 5.22.1.
61. Richlin, "Reading Ovid's Rapes," pp. 162, 163, 165.
62. Louise Pearson-Smith, "Audience Response to Rape: Chaerea in Terence's *Eunuchus*," *Helios* 21 (1994): 21–38.
63. Zola M. Packman, "Call It Rape: A Motif in Roman Comedy and Its Suppression in English-Speaking Publications," *Helios* 20 (1993): 42–55. See also Richard P. Saller, "The Social Dynamics of Consent to Marriage and Sexual Relations: The Evidence of Roman Comedy," in *Consent and Coercion*, pp. 83–104. For criticism of the aesthetic sanitization of "heroic" rape, see Diane Wolfthal, *Images of Rape: The "Heroic" Tradition and Its Alternatives* (Cambridge: Cambridge University Press, 1999), pp. 1–35.
64. Terence, *Eunuchus* lines 645, 647, 659; trans., p. 387.
65. Patricia Watson, "*Puella* and *virgo*," *Glotta: Zeitschrift für griechische et lateinische Sprach* 61 (1983): 120–23. Terence, *Eunuchus* lines 645–46, cf. 820.
66. Terence, *Eunuchus* lines 1035–36, 1047.
67. Marjorie O'Rourke Boyle, "Augustine in the Garden of Zeus: Lust, Love, and Language," *Harvard Theological Review* 83 (1990): 117–39.
68. *Confessions* 1.16.26. For citations from memory, see James J. O'Donnell, "Augustine's Classical Readings," *Recherches augustiniennes* 15 (1980): 144–75. For the philology, see Elaine Fantham, "*Stuprum*: Public Attitudes and Penalties for Sexual Offences in Republican Rome," *Échos du monde classique/Classical Views*, 36 (1991): 267–91.
69. Terence, *Eunuchus* line 1.
70. See Thomas Habinek, *The World of Roman Song: From Ritualized Speech to Social Order* (Baltimore: The Johns Hopkins University Press, 2005), pp. 120–21, 113–14, 110, 118, 114–15.
71. Boyle, *Divine Domesticity*, p. 10.
72. Ovid, *Metamorphoses* 6.113, 11.117.
73. Augustine, *De civitate Dei* 2.7, ed. Bernard Dombart and Alphonse Kalb, 2 vols. (Turnhout: Brepols), 1981). Citing Terence, *Eunuchus* 645–46, 648, 659; trans., p. 387. 2.12.

74. Augustine, *Epistolae* 91.4–5, in *Patrologiae cursus completus, series latina*, ed. J.-P. Migne, 221 vols. (Paris, 1800–75), 33: col. 315. For abhorrence of the theatrical gods, see also Sabine MacCormack, *The Shadows of Poetry: Vergil in the Mind of Augustine* (Berkeley and Los Angeles: University of California Press, 1998), pp. 198–99, 212.
75. *Confessions* 2.6.14, 2.6.12, 2.2.2, 2.2.3.
76. Catharine Edwards, *The Politics of Immorality in Ancient Rome* (Cambridge: Cambridge University Press, 1993), pp. 173, 175.
77. See J. Reginald O'Donnell, "The Meaning of *silva* in the Commentary on the *Timaeus* of Plato by Chalcidius," *Mediaeval Studies* 7 (1945): 8–9.
78. *Confessiones* 2.2.4, 1.16.26.
79. Aelius Donatus, *Commentum Terentii* at lines 584–85.
80. *Confessiones* 2.3.5, 2.6.13, 8.12.28, cf. Terence, *Eunuchus* line 585.
81. For self-control, see Edwards, *Politics of Immorality*, p. 5.
82. Terence, *Eunuchus* line 550. For the rarity, thus impressiveness, of oaths by Jupiter in Roman comedy, see Barsby, ed., *Eunuchus*, p. 189.
83. *Confessiones* 3.2.2. Donnalea Dos, *The Idea of the Theater in Latin Christian Thought: Augustine to the Fourteenth Century* (Ann Arbor: University of Michigan Press, 2004), pp. 12–29; Edwards, *Politics of Immorality*, pp. 98–136.
84. Augustine's citation of Terence adds "in pariete" in deference to the fashion of Roman murals, rather than Greek panels. For that art, see Roger Ling, *Roman Painting* (Cambridge: Cambridge University Press, 1991), pp. 1–11; and for their depiction of myth and history, pp. 101–41. This is the sole reference in Terence to a painting. Charles Knapp, "References to Painting in Plautus and Terence," *Classical Philology* 12 (1917): 153–54. For the commonplace, see Barsby, ed., Eunuchus, p. 198. For erotic art as motivational, see Molly Myerowitz, "The Domestication of Desire: Ovid's *parva tabella* and the Theater of Love," in *Pornography and Representation*, p. 137.
85. Augustine, *De trinitate libri XV* 11.2.5, ed. W. J. Mountain, 2 vols. (Turnhout: Brepols, 1968). He probably knew this from the physician Vindicianus, the author of *Gynaecia*. He appears as a friendly advisor to Augustine in *Confessiones* 4.3.5–6. See also Augustine, *Epistulae* 138.3, in *Patrologia latina*, 33: col. 526. See Vindicianus, *Gynaecia*, in *Opera*, in *Priscianus "Euporiston": Accedunt Vindicianus Afri quae feruntur reliquiae*, ed. Valentin Rose (Leipzig: Teubner, 1894), pp. 426–66. Vindicianus is the more probable source for the anecdote about conception than Pierre Courcelle's opinion that Augustine read Soranus, *Gynaecia* 1.39 in the Greek. Courcelle, *Later Latin Writers and Their Greek Sources*, trans. Harry E. Wedeck (Cambridge, Mass.: Harvard University Press, 1969), pp. 195–96.

86. For the ideal, see Karl Kilinski II, "Greek Masculine Prowess in the Manifestations of Zeus," in *Myth, Sexuality, and Power: Images of Jupiter in Western Art*, ed. Frances V. Keuren (Providence, R.I.: Brown University Center for Old World Archaeology and Art; Departement d'Archeologie et d'histoire de l'art, College Erasme, Louvain-La-Neuve, Belgium, 1998), pp. 29–50. See also Boyle, "Augustine in the Garden of Zeus."
87. See Bonner, *Roman Declamation*, p. 90.
88. Terence, *Eunuchus* line 859.
89. F. van der Meer, *Augustine the Bishop: The Life and Work of a Father of the Church*, trans. Brian Battershaw and G. R. Lamb (London: Sheed & Ward, 1961), pp. 135, 181, citing Augustine, *Sermones* 224.3, in *Patrologia latina*, 45; Gervase Corcoran, *Saint Augustine on Slavery* (Rome: Institutum Patristicum "Augustinianum," 1985), pp. 10, 29–30.
90. Romans 5:14–15. Augustine used the Vetus Latina (Old Latin) Bible for all Pauline citations. H. A. G. Houghton, *Augustine's Text of John: Patristic Citations and Latin Bible Manuscripts* (Oxford: Oxford University Press, 2008), p. 84. However, for this verse the phrase is identical to the Vulgate.
91. *Confessiones* 8.12.29, reading Rom. 13:13–14.
92. Augustine, *Expositio quarumdam propositionum ex Epistola ad Romanos*, in *Patrologia latina*, 35:2063, with further reference to 1 John 3:20. See also *Confessiones* 4.19.14. Augustine, *Epistolae ad Romanos inchoata expositio*, in *Patrologia latina*, 35:2087–2106, treats only the first chapter.
93. *Confessiones* 5.3.3, 5.
94. Augustine, *Contra Faustum* 19.2, ed. Joseph Zycha (Vienna: F. Tempsky, 1891), p. 497.
95. *Confessiones* 15.26.
96. Ibid. 7. 9.14.
97. Aristotle, *Categoriae* 5 2a–4a.
98. Ibid. 4.16.28–29.
99. Ibid. 2.4.9.
100. Without reference to Augustine, see Edwards, *Politics of Immorality*, p. 4. See also Wolfgang Blösel, "Die Geschicthe des Begriffes *mos maiorum* van den Anfängen bis zu Cicero," pp. 25–97, in *Mos maiorum: Untersuchungen zu den Formen der Identitätsstiftung und Stabilisierung in der römischen Republik*, ed. Bernard Linke and Michael Stemmler (Stuttgart: Franz Steiner, 2000).
101. *Confessiones* 2.3.8, 2.4.9.
102. Ibid. 2.4.9.
103. Terence, *Eunuchus* lines 383–87.
104. *Confessiones* 2.3.7, 2.8.16. 2.2.9, cf. Terence, *Eunuchus* line 839. *Confessiones* 2.9.17, 2.8.16.

105. Aristotle, *Ethica Eudemia* 7.1.7 1235a.
106. Erasmus, *Adagia*, in *Opera omnia*, ed. Jean Leclerq, 11 vols. (Leiden: Petrus van der Aa, 1703–6), 2:509; trans. R. A. B. Mynors, in *The Collected Works of Erasmus* (Toronto: University of Toronto Press, 1974–), 33:169–70. *Adagia*, ed. M. L. van Poll-van de Lisdonk et al., in *Opera omnia* (Amsterdam: North-Holland, 1971): 2–1:236–38; trans. Margaret Mann Phillips, *Collected Works of Erasmus*, 31:165–66. See also *Adagia*, pp. 38–42. See also Kathy Eden, *Friends Hold All Things in Common: Tradition, Intellectual Property, and the "Adages" of Erasmus* (New Haven, Conn.: Yale University Press, 2001).
107. Michael Pakaluk, "The Egalitarianism of the *Eudemian Ethics*," *Classical Quarterly* 48 (1998): 411, 423–24. See also Anne Marie Dziob, "Aristotle's Friendship: Self-Love and Moral Rivalry," *Review of Metaphysics* 46 (1993): 781–801; Alex J. London, "Moral Knowledge and the Acquisition of Virtue in Aristotle's *Nicomachean* and *Eudemian Ethics*," *Review of Metaphysics* 54 (2001): 553–83.
108. Aristotle, *Ethica Eudemia* 7.1.7 1235a.
109. Cicero, *De officiis* 2.39–40; *On Duties*, trans. M. T. Griffin and E. M. Atkins (Cambridge: Cambridge University Press, 1991), pp. 77–78. Against theft, see also 3.5.21.
110. Andrew R. Dyck, *A Commentary on Cicero, "De officiis"* (Ann Arbor: University of Michigan Press, 1996), p. 418, who suggests Panaetius's inspiration is Plato, *Respublica* 351c. See also Aristotle, *Rhetorica* 1397b.
111. See Bonner, *Education*, pp. 261–62, 266.
112. Cicero, *De re publica* 3.22 is frequently cited. For Jupiter, see Cicero, *De legibus* 2.4.10.
113. Brent D. Shaw, "Bandits in the Roman Empire," *Past and Present* 105 (1984): 4; Nicholas K. Rauh, *Merchants, Sailors, and Pirates in the Roman World* (Stroud, Gloucestershire: Tempus, 2003), pp. 169–77.
114. Cicero, *De officiis* 3.11.49.
115. Cicero, *Epistulae ad Atticum* 5.20. For gubernatorial responsibility, see Shaw, "Banditry," pp. 14, 19.
116. For Cilician piracy, see Rauh, *Merchants, Sailors, and Pirates*, pp. 169–200; Philip de Souza, *Piracy in the Graeco-Roman World* (Cambridge: Cambridge University Press, 1999), pp. 97–148. For revolts in Cilicia still in the fourth century C.E., see Keith Hopwood, "Bandits, Elites and Rural Order," in *Patronage in Ancient Society*, ed. Andrew Wallace-Hadrill (London: Routledge, 1990), p. 173.
117. de Souza, *Piracy*, pp. 135–36, 149–57, with citation of Cicero, *In 2 Verrem* 5.76 at p. 156.
118. Augustine, *De magistro* 5.16, ed. K.-D. Daur, in *Contra academicos, De beata vita, De ordine, De magistro, De libero arbitrio*, ed. W. M. Green et al. (Turnhout: Brepols, 1970).

119. Cicero, *De officiis* 3.107; *On Duties*, trans., p. 141.
120. de Souza, *Piracy*, p. 132.
121. Rauh, *Merchants, Sailors, and Pirates*, pp. 194–95.
122. See above, pp. 49–50.
123. Rauh, *Merchants, Sailors, and Pirates*, pp. 196–97.
124. Shaw, "Bandits," pp. 10–12, 8–9, 22, 50, 6–7, 24–27, 19, 23. For the vocabulary, see also de Souza, pp. 12–13. For outsiders, see also Ramsay MacMullen, *Enemies of the Roman Order: Treason, Unrest, and Alienation in the Empire* (Cambridge, Mass.: Harvard University Press, 1966), pp. 192–219.
125. Rauh, *Merchants, Sailors, and Pirates*, pp. 165, 168, 187, 189.
126. Shaw, "Bandits," pp. 20–21.
127. *De civitate Dei* 4.4. Cited by Shaw, "Bandits," p. 3; Rauh, *Merchants, Sailors, and Pirates*, p. 195. By this date Augustine's knowledge of their habits may owe also to Apuleius, *Metamorphoses* 3.27–4.22, with sharing the loot at 3.28 and disguises at 4.14–15, 7.8. For the vocabulary, see Werner Riess, *Apuleius und die Räuber: Ein Beitrag zur historischen Kriminalitätsforschung* (Stuttgart: Franz Steiner, 2001), pp. 32–44. See also Vincent Hunink, "'Apuleius, qui nobis Afris Aer est notior': Augustine's Polemic against Apuleius in *De civitate Dei*," *Scholia: Studies in Classical Antiquity* 12 (2003): 82–95. For other fictional authors on the pirate share, see de Souza, *Piracy*, p. 216. For piracy in Homer, whom Augustine read in school, see pp. 17–21.
128. *De civitate Dei* 4.1.
129. See de Souza, *Piracy*, p. 199.
130. Plutarch, *Pompey* 24.3; *Vitae*, trans. Bernadotte Perrin, 11 vols. (London: William Heinemann, 1914–26), 5:175. Rauh, *Merchants, Sailors, and Pirates*, p. 198. See also N. R. E. Fisher, *"Hybris": A Study in the Values of Honour and Shame in Ancient Greece* (Warminster, Wiltshire: Aris and Phillips, 1992), pp. 156–60, 162, 164, 178.
131. *Confessiones* 2.6.12, 2.6.14, 2.4.9.
132. See Terence, *Eunuchus* line 326; trans., p. 349; ed. Barsby, p. 147.
133. Henry A. Ormerod, *Piracy in the Ancient World: An Essay on Mediterranean History* (Liverpool: Liverpool University Press, 1978), pp. 68–72.
134. E.g., Miguel de Cervantes, *Don Quijote*, part two, chapter 60, cited by the *Oxford Dictionary of English Proverbs*, 1948 ed., s.v.
135. Ormerod, *Piracy in the Ancient World*, pp. 13–14, 207; Shaw, "Banditry," p. 39.
136. Possidius, *Vita* 1, in *Patrologia latina*, 32. Geoffrey Rickman, *The Corn Supply of Ancient Rome* (Oxford: Clarendon, 1980), pp. 13, 50–52, 67–71, 112, 167–69, 204–6, 231–35.

137. de Souza, *Piracy*, pp. 213, 214, 224, 229, 231.
138. *Confessiones* 7.21.27, 5.8.15. Othmar Perler with Jean-Louis Maier, *Les voyages de saint Augustin* (Paris: Études augustiniennes, 1969), pp. 52–53.
139. *Confessiones* 3.8.16, 7. 21.17.
140. See James Yates, "Latrunculi," in *A Dictionary of Greek and Roman Antiquities*, ed. William Smith (London: John Murray, 1875), p. 670, to which may be added the earliest reference, Varro, *De lingua latina* 10.2.
141. Shaw, "Banditry," p. 9; Hopwood, "Bandits, Elites and Rural Order," p. 179.
142. Bonner, *Declamation*, pp. vi–vii, 6–8, 133; Ormerod, *Piracy*, pp. 264–66.
143. Ormerod, *Piracy*, pp. 260–64.
144. Terence, *Eunuchus* lines 108–18.
145. Without reference to Augustine, see Rauh, *Merchants, Sailors, and Pirates*, p. 197; de Souza, *Piracy*, pp. 215–16.
146. Without reference to Augustine, see Julian Pitt-Rivers, "Honour and Social Status," in *Honour and Shame: The Values of Mediterranean Society*, ed. J. G. Peristiany (Chicago: University of Chicago Press, 1974), pp. 11, 21, 23, 31, 35–36. See also Carlin A. Barton, *Roman Honor: The Fire in the Bones* (Berkeley and Los Angeles: University of California Press, 2001).
147. See note 13.
148. *Confessiones* 2.4.9.
149. Joseph Heinrich, "Cooperation, Punishment, and the Evolution of Human Institutions," *Science* 312 (2006): 60–61.
150. William M. Green, *"Initium omnis peccati superbia": Augustine on Pride as the Original Sin* (Berkeley: University of California Press, 1949).

Aquinas's Law of the Heart: Natural Reason

Thomas Aquinas's *Summa theologiae* premised a natural law of the heart on the providential imprint in creatures of a share in the Creator's eternal law. A human, as a rational creature provident for self and others, shared in his eternal law superlatively. As Aquinas defined that human share, "Whence it itself participates in the eternal law, through which it has an inclination to the just act and end. And such participation of the eternal law in the rational creature is called the natural law." Aquinas developed his scholastic question in six articles: "whether the natural law is a habit, whether the natural law contains many precepts or one only, whether all virtuous acts are from the natural law, whether the natural law is one for everybody, whether the natural law can be changed," and "whether the natural law can be abolished from the human heart." Aquinas argued against the abolition of the natural law from the heart by altering a sentence of Augustine's *Confessions* from its rhetoric to his own reasoning. "But to the contrary is what Augustine says, 'Thy law has been written in human hearts, which not even injustice itself deletes.' But the law written in human hearts is the natural law. Therefore, the natural law cannot be deleted."[1]

Aquinas did not demonstrate his minor premise that "the law written in human hearts is the natural law." He cited partially a dependent clause of the Epistle to the Romans 2:14: "When the nations, who do not have the law, naturally do those things that belong to the law." He named his source as the *Glossa ordinaria*, a twelfth-century collection of comments

M. O'R. Boyle, *Cultural Anatomies of the Heart in Aristotle, Augustine, Aquinas, Calvin, and Harvey*,
https://doi.org/10.1007/978-3-319-93653-6_3

on the Vulgate Bible that was in standard usage. Aquinas cited closely, "Although they do not have the written law, nevertheless they have the natural law, whereby everyone perceives and is aware of what is good and what is evil."[2] He did not, however, add "naturally" (*naturaliter*) as meaning "natural reason," which the *Glossa* identified as Origen's interpretation. As the *Glossa* stated, "That is, illuminated by natural reason they discern the things the law makes for doing and avoiding." Origen's commentary on Romans had specified "what they can sense naturally" and "this natural power of discernment."[3] Nor did Aquinas reference the *Glossa* for the "heart" as faith operating "in the innermost affect" and "through love," or differently so as "firmly fixed in reason."[4] Yet, Aquinas's law of the heart was natural reason—until near the end of his life his *De motu cordis* extracted the human rational soul from the heart. His psychology of the soul as the form of the body ultimately denied the natural law residence in the heart even temporarily. Yet, his theory of the natural law of the heart in *Summa theologiae* prevailed historically to the modern ethics in his tradition.

Aquinas's major premise in *Summa theologiae* for the permanence of the natural law of the heart altered the referenced sentence of Augustine's *Confessions* by the omission of twelve words. Those deletions reordered its syntax, thus its meaning. Augustine wrote: *Furtum certe punit lex tua, domine, et lex scripta in cordibus hominum, quam ne ipsa quidem delet iniquitas: quis enim fur aequo animo furem patitur.* "Thy law, o Lord, certainly punishes theft, and the law written in human hearts, which even injustice itself does not delete, for indeed what thief suffers a thief with an even mind."[5] Aquinas's *Summa theologiae* rendered it as: *lex tua scripta est in cordibus hominum, quam nec ulla quidem delet iniquitas.* "Thy law is written in human hearts, which not even injustice itself can delete."[6] Aquinas deleted Augustine's essential context, his adolescent theft of pears, the misdeed he was recollecting with a profound examination of his sinful motives. The prime position of *furtum* in Augustine's word order emphasized "theft" as his subject and his extensive meditation on that theft confirmed his context. Aquinas further omitted the apostrophe, *domine* "o Lord," which served as the focal middle term of Augustine's sentence. It invoked and identified God as the praiseworthy object of his address about his theft. That apostrophe, *domine* "o Lord," also separated syntactically the two distinct subjects of Augustine's verb *punit* "punishes." Aquinas collapsed the middle of Augustine's sentence to equate "thy law" with "the law written in human hearts." However, Augustine had consistently differentiated between "thy law" as the written Bible from "the law written in human hearts" as human custom—hyperbolically as outlawry.[7]

Augustine's *Confessions* was epideictic rhetoric,[8] the classical genre for persuading praise or blame of an individual, such as God or Augustine. That genre, in a range from panegyric to invective, did not prove universal abstract truth and falsehood but persuaded an audience toward a judgment of its particular subject.[9] Augustine composed his *Confessions* "to praise the just and good God, and to excite human understanding and desire toward him."[10] The objective of Aquinas's *Summa theologiae* was not such a rhetorical arousal; it was not, in Augustine's verb, "exciting." As Aquinas began *Summa theologiae*, "The plan of our intention in this work is to treat the matters that pertain to the Christian religion in a manner conforming to the education of beginners."[11] Their respective argumentation differed fundamentally from rhetoric to logic. Cicero's classical *De inventione*, a text Aquinas knew, explicated that rhetorical arguments derived from the first category, invention.[12] Augustine, a professional rhetor, invented his epideictic *Confessions* topically, "from evils and from goods."[13] The epideictic topics, or argumentative places, were the attributes of persons—in mind, body, and circumstances—with the end of the honorable, or virtue. Those personal attributes were name, nature, manner of life, fortune, habit, feeling, interests, purposes, achievements, accidents, and speeches.[14] To ignore the propriety of place was to violate decorum, to be, as Cicero judged, "tactless."[15] Augustine invented from those argumentative places topically, "from goods and evils," many and particular, toward divine praise and human blame. The medieval philosopher Boethius in *De topicis differentiis*, which Aquinas cited, neatly differentiated between dialectic, which is "restricted to question and answer," and rhetoric, which discourses continuously on a subject. "The dialectical discipline examines the thesis only; a thesis is a question not involved in circumstances. The rhetorical [discipline], on the other hand, investigates and discusses hypotheses, that is, questions hedged in by a multitude of circumstances. Circumstances are who, what, where, when, why, how, and by what means."[16] Aquinas identified his scholastic question on the natural law with dialectic, citing Boethius in it on "propositions commonly known in themselves by everyone."[17] However, Aquinas's proof text for the permanence of the natural law in the heart was adapted from rhetoric, not dialectic.

Aquinas appealed to Augustine's personal testimony but adapted it for a universal statement. Aquinas framed his question from his response to it. Q. "Whether the law of nature can be abolished from the heart of men." Against objections to its permanence he decided. A. "But to the contrary is what Augustine says, 'Thy law has been written in the hearts of men, which not even injustice itself can delete.'"[18] Aquinas's question about the deletion of the natural law was thus posed before the sentence from which it was derived. However, Augustine's topical argument "from evils" followed from his preceding sentence, which concluded emphatically with "my injustice (*iniquitas mea*)." That blame agreed with the purpose of epideictic rhetoric to praise or blame an individual. However, Augustine's next sentence, about "the law written in human hearts, which even injustice itself (*ipsa iniquitas*) does not delete,"[19] was not a logical inference from the particular to the general. His "injustice itself" was projected from his basic topic of his personal evils to recall by chiasmus the effect of *iniquitas mea*, "my injustice." His amplification was hyperbolic, "for indeed what thief suffers a thief with an even mind." That emphasis cited proverbial thieves' honor, a social sympathy of like to like that was neither divine nor natural law. In Roman civil law it was outlawry.[20]

The Roman realities of bandits on land and pirates at sea Aquinas abstracted to rule banditry as being against the natural law. Although he had fragments of textual evidence toward understanding Augustine's rhetoric, his own method was logical. Augustine in *De civitate Dei* stated his knowledge of thieves' honor in sharing bounty. "With justice removed," he argued, "what are kingdoms but huge banditries? Because what are banditries but tiny kingdoms? The legal power of the men is governed by the chief's command and bound by a social pact, and the loot is divided by an agreeable law." Augustine judged the pirate chief who marauded in a flimsy ship no different from Alexander the Great whose mighty armies warred on the entire world.[21] Aquinas cited that very text in *Summa theologiae* but conflated robbery (*rapina*) with banditry (*latrocinium*).[22] Augustine's context, which was Roman law, distinguished them, with robbery against the law but banditry outside the law.[23]

Augustine introduced bandits in *De civitate Dei* with his personal attendance at the theater, where the fictional crimes of the pagan gods were performed "to be imitated as if by divine authority."[24] His *Confessions* had formidably and extensively blamed the rhetorical exemplarity of the theater for godly permission to sin. He identified himself as an adolescent thief of pears then thrown to pigs with Chaerea, the protagonist of Terence's comedic *Eunuchus*, who stole a girl's virginity—pears being a classical symbol for sexual maturity, pigs being classical slang for the female genitals.[25] As a schoolboy at the Benedictine abbey of Monte Cassino, Aquinas would have read Terence's comedies to learn colloquial Latin.[26] He cited Terence's *Eunuchus* in conformity with Aristotle's definition of comedy as presenting inferior characters who are laughable as base or ugly.[27] Aquinas's citations explicated the vices of imprudence and lust. He quoted *Eunuchus* three times in *Summa theologiae* and once in his disputed question *De malo*. In arguing about imprudence in *Summa theologiae* he accorded Terence's comedy the authority of Aristotle's *Ethica nicomachea*, which he cited equally in that question. Aquinas reasoned that lust results in duplicity as mental fluctuation, quoting from *Eunuchus* "in love is war, and back again peace and truce."[28] Aquinas's citation changed, by the omission of seven words and the transposition of four, the speech on the vagaries of love by the slave Parmeno to his confused master, Phaedria. Terence wrote, "Master, when a thing has no logic to it and no means of control, you can't rule it by logic. A love affair has all these symptoms: wrongs, suspicions, quarrels, truces, war, peace again. If you try to impose certainty on uncertainty by reason, you'd achieve no more than if you set about going insane by reason."[29] In reasoning about lust Aquinas repeated the slave's opening line about "lecherous love" as evidence that a concupiscent man could be hindered from acting rationally. "Hence, Terence of the man who declared he would leave his mistress 'One little false tear will undo these words.'"[30] However, the comedic plot did not turn on such a moral decision. The lover had not declared that he would leave his mistress, only that he vacillated between anger and desire toward her. Also, different from Aquinas's assumption, the mistress toying with the lover's emotions had shut him out to please a rival. Aquinas's citation of the play then continued the slave's speech, which mimicked the amatory indecision of his master and warned him against seduction. "And as for your present angry thoughts—'I—her? When she—him? When she—me? When she won't—? Just let it be, I'd prefer to die, she shall realize what sort of man I am'—god knows she'll quell that sort of talk with one tiny little false tear, which she's just managed to squeeze out by rubbing her eyes all pathetically." The lover lets her have her way (*mos gerundust*), in the indulgence of comedy,[31] the deference to human custom (*mos*) that Augustine's *Confessions* villainized.[32] In *De malo* Aquinas rehearsed the slave's lines to philosophize that lust eradicates the deliberation necessary to rational acts as directive of human acts.[33] All of his citations were from act one scene one of *Eunuchus*; indeed all were from the same speech. Perhaps it was a familiar set piece for memorization at school. Aquinas would not have considered Augustine's theft of the pears tossed to pigs as metaphorical for rape because he distinguished spiritual sins from carnal sins, with theft distinct from fornication.[34]

Aquinas, as a Christian teacher, argued from authority with Augustine as his prime traditional source. Aquinas quoted his *Confessions* forty-two times, three times already from book two with the sentence he adapted to prove the permanence of the natural law of the heart. He then quoted that same book four times more. Although two of his citations were accurate, and two differed in minor respects of semantics or order, one was half invented, and one exhibited a liberty of omission. In one question about fear he accurately cited Augustine: *timor insolita et repentina exhorrescit, rebus quae amantur adversantia, dum praecavet securitati.* In another question about fear he abridged and reordered that sentence as: *timor securitati praecavet.* All of those citations belonged to Augustine's personal examination of conscience about his theft of the pears, precisely the contextual *furtum* that Aquinas deleted to posit an abstracted natural law. Three of those citations involved Augustine's essential blame of exemplarity and imitation—how human fear, wrath, and pride mimicked the divine attributes. Aquinas abstracted the citations from Augustine's mimetic context to conscript them as definitions of human appetites in his scholastic questions: "Whether daring is contrary to fear?" "Whether the object of anger is good or evil?" and "Whether pride is a sin?"[35] Aquinas thus appropriated Augustine's rhetorical *Confessions* as a proof text for his own philosophical theory of the natural law of the heart.

Nature

In review, Aquinas premised his theory of natural law in *Summa theologiae* by affirming that every creature governed by divine providence was imprinted to share in its eternal law. A human, as a provident rational creature, shared in that eternal law supremely. "Whence it itself participates in the eternal law, through which it has an inclination to the just act and end. And such participation of the eternal law in the rational creature is called the natural law."[36] Natural law belonged to Aquinas's treatment of appetite, concerning the passions of the soul, the will, its objects of good and evil, habits, virtues and vices, the gifts of the Holy Spirit, sin, and law. He introduced the law according to the extrinsic principles of acts: the devil for evil by temptation, but God for good by law and grace.[37] Aquinas presented the law as an extrinsic principle of divine motivation to good and from evil. As he defined, "Law is a certain rule and measure of acts, whereby someone is induced to acting, or restrained from acting; for 'law' (*lex*) is named from binding (*ligare*) because it obligates toward acting." This obligation Aquinas derived from rational command. By a presupposed act of the will in accord with reason, reason commanded the execution of the act. Aquinas conformed to his prior acceptance of Aristotle's psychology in which reason was superior to will. As the rule and measure of human acts, Aquinas declared the law rational as their first motive principle.[38]

Aquinas then reviewed the types of law, which he classified as eternal, natural, human, old, and new.[39] On the eternal law, he defined law more specifically. "Law is nothing else than some precept of practical reason in the leader who governs a particular perfect community." As he argued, since the universe was ruled by divine providence, it was ruled by divine reason, which was eternal; therefore, its law was eternal. Aquinas considered whether the eternal law rendered the natural law superfluous. He asked whether, since irrational animals acted by natural appetite, whereas humans acted by reason and will, there was consequently no natural law for humans. His response divided the eternal law twofold. It was in the ruler and measurer—God—absolutely, and in the ruled and measured—creatures—participatively. For creatures, "All things participate equally in the eternal law, insofar as obviously from its impression they have inclinations to proper acts and ends." Yet a rational creature, a human, was more excellently subject to divine providence insofar as he himself shared in it actively by providing for himself and for others. "Whence also he participates in the eternal reason by which he has the inclination to the owed act and end. And such participation of the eternal law in the rational creature is called the natural law."[40]

Although whether or not Aristotle theorized a natural law is debated,[41] his ethics and metaphysics have been referenced commonly to interpret Aquinas's natural law. Aristotle's physics has been neglected. Yet, before Aquinas resumed late in 1270 the first part of the second part of *Summa theologiae*, with its question on natural law, he had composed in 1268–1269 his Aristotelian commentary *In libros Physicorum*.[42] His further Aristotelian writings *Sententia libri ethicorum* and *Sententia libri Metaphysicae* dated to after his question on natural law. Although he was hardly ignorant of those works of Aristotle, and had begun around 1270 an index to the two on ethics,[43] Aquinas had not yet commented on them as methodically. In his question on natural law in *Summa theologiae* Aquinas's cited three Aristotelian works—*Ethica*, *Metaphysica*, and *Physica*—the last on the rational process from the general to the particular.[44] Yet, Aristotle's *Physica* had fundamentally defined "nature" as Aquinas cited it. Aquinas's deliberations on its problematic meaning for a Christian philosopher occasioned his own definition of natural law as a "participation."

An incoherence to Aquinas's "natural law" was his definition of law as extrinsic and his predication of nature as intrinsic. As a volitional principle, the law involved movements. A natural motive principle, innate in in the soul, belonged to the science of motion, or physics. Aquinas had established human movement on Aristotle's natural philosophy. Animal life commenced and continued by self-movement. To be natural was to possess a motive principle intrinsically. As Aquinas explained, "In every bodily nature the more perfect are the living bodies, when the noun 'nature' itself has been transferred from living things to all natural things." As he repeated Aristotle's explanation, "nature" first signified the generation of living things, or birth. Then, "Because living things are generated from a conjoined principle—fruit from a tree, a fetus from a mother to whom it is bonded, it followed that the noun 'nature' has been extended to every principle of movement that exists in what is moved."[45] The natural law was thus natural by Aristotle's definition because it possessed an *intrinsic* principle of self-movement. However, Aquinas presented all law as an *extrinsic* principle of action originating in God as the first mover. There were distinct basic principles of movement, extrinsic and intrinsic. Extrinsic motion pushed and pulled a body from without, intrinsic motion pushed and pulled a body from within. For living things, precisely as alive, their bodies were informed by soul: plants with a vegetative soul, all animals with also a sensitive soul, and humans alone with also an intellectual soul. Those differences challenged Aquinas toward a more complex explanation of movement than Aristotle had ventured. Aristotle's natural philosophy focused on the common denominator of all animals, the quiddity of animal. Aquinas's *Summa theologiae* focused on the privileged difference of human animals, the quiddity of human.

Aquinas had originally from Aristotle defined the natural by *possession* of an *intrinsic* principle of movement. In Aquinas's new definition the natural became *participation* in an *extrinsic* principle of movement.[46] Law of all types Aquinas introduced as extrinsic principles of actions. How, then, could the eternal law as an extrinsic principle of action become an intrinsic principle of action as the natural law? His theological difficulty was that Aristotle's nature existed in an eternal uncreated universe, whereas Aquinas's nature existed in a temporal created universe. Aquinas tried to obviate the difference, which posed the contradiction between intrinsic and extrinsic principles. In his definition of natural law he posited that the extrinsic principle—supremely the eternal law in God—was divinely imprinted (*ex impressione*) in humans as an intrinsic principle. Thus it was natural as sharing that extrinsic principle. The divine imprint went unexplained, however, except for his biblical citation, "The light of your countenance is marked upon us" (Ps. 4:7 Vulg.).[47] In his commentary on that psalm in his final year[48] Aquinas offered that "natural reason set in us teaches (*docet*) us to discern good from evil." The divine countenance meant "that by which we know God, just as a man is known by his face; that is, the truth of God. From this truth of God shines a likeness of his light in our souls. And this is like a light, and is signed upon it, because it is superior in us; and it is like some sign upon our face, and by this light we are able to know the good."[49]

Aquinas resumed in *Summa theologiae* his argument that the light of natural reason, which was the function of the natural law, was nothing else than an imprint on humans of the divine light. He concluded with his definition, "Therefore, the natural law is the rational creature's participation in the divine law."[50] Aquinas did not argue or establish philosophically his premise of a divine imprint on human reason toward his conclusion of a natural law. He cited Scripture, as consistent with the premise of his *Summa theologiae* that "it was necessary for human salvation that there should be some doctrine according to divine revelation, beyond the philosophical disciplines, which investigate by human reason."[51] However, his *Summa theologiae* mistook his scriptural proof text as a divine imprint on human reason. Aquinas was disadvantaged by his ignorance of the Hebrew Masoretic Text, dating from the sixth to the tenth centuries, which had *nesāh* from *nasa/nasah* ("lift up"). He depended on the Vulgate Latin translation that circulated in uncritical manuscripts.[52] It had *signatum est* ("is marked"). The accurate and the traditional Hebrew verse is not Aquinas's premise, "The light of your countenance is marked on us" (Ps. 4:7 Vulg.) but, rather, the petition "Lift up the light of your countenance upon us" (v. 6 RSV, since the AV).[53] The verse continued "You have given gladness (*laetitia*) in my heart"—not reason (*ratio*) in the heart, as Aquinas argued for the natural law. Aquinas borrowed from Augustine's psalm commentary a simile of the human mind impressed to the divine image like a ruler's effigy minted on a silver coin. Augustine had regarded the Septuagint translation of the Hebrew into Greek as authoritative and he preferred the Itala, the Vetus latina (Old Latin Bible), for its Latin translation.[54] For that verse the Vulgate *signatum est* and the Septuagint *esēmeiōthē* (LXX) agreed. Most sources for the Vetus latina had *signatum est*, but two codices with *exalta* ("raise") and *leva* ("lift up") were faithful to the Hebrew *nesāh*.[55]

Augustine explicated and Aquinas copied *signatum est*, "'stamped on us' as a denarius is stamped (*signatur*) with the king's image."[56] However, in medieval numismatics the face of the temporal authority on a coin did not make it money. Its intrinsic value of precious silver and its nominal value by a fixed rate made a coin a measure of values and item of exchange. Nor were coins pressed one from another; all issued from a common die.[57] Coinage was nevertheless Aquinas's simile for the divinely imprinted human mind—at least the male mind. His proem to the first part of the second part of *Summa theologiae* cited John Damascene's *De fide orthodoxa* on the divine image in humans. It meant "the intellect and freedom of choice *per se* denoting power." Aquinas intended to treat that image as the principle and power of human works.[58] He stated that both sexes were created in the divine image equally by their intellectual nature. However, he judged that females were created functionally "second-rate" (*secondario*) to males. Females were "naturally subject to men because in males the discernment of reason abounded more." Aquinas declared that "the man is the principle and end of the woman just as God is the principle and end of all creatures." Woman was practically, if not ideally, only a copy of the divine image in man. Through participation in male rationality she participated indirectly in God's eternal law as the natural law.[59] Aquinas's natural subjection of females to males he precisely stated as "economic or civil," meaning "for utility and good."[60] That argument again evidenced his equivocation on the meaning of nature. For, although females were created in the divine image by their natural intellectuality, they were subject to the male image of intellectuality by their inferior rationality. Yet, that economy was as natural as intellectuality was natural. His gendered argument thus weakened his concept of the natural law. For Aquinas, the natural law belonged to the practical reason. Yet, because of their inferior rationality, actual female knowledge of and observance of the natural law was naturally feeble. Women were naturally subject to men's superior rational discernment of the natural law, as of every law. Practically—and the natural law concerned practical reason toward practical ends—females participated in the natural law only indirectly through males, who alone participated directly in the eternal law. Aquinas's belief in the rational inferiority of females reduced the direct knowability of the natural law by arguably half of humans.

For his proposed biblical proofs for the divine imprint on humans as reason, Aquinas distinguished internal from external principles of movement. It belonged to the mover as cause "to imprint the form, to dispose to the form, and to give the motion consequent on the form." He stated that if something had no notion of its end, it still had within it a principle of action or movement. But, it did not act or move from itself on account of the end that was a principle in itself, but from the principle that imprinted its movement to the end. It was, therefore, not self-moving but moved. Only those who knew the end were self-moving to the end.[61] However, the divine light imprinted on reason as the natural law was not innate to reason. That principle, the natural law, was derived, imputed, and participatory. What meaning did "natural," as the principle of intrinsic movement, retain?

Aquinas posited law as intrinsic, thus after Aristotle natural to humans. Yet, he did so by reference to its extrinsic mover, the Governor, as intrinsic to the human soul through rational participation of his eternal law. Aquinas's reasoning involved his opinion on naming. As he argued basically on the good, "Everything whatsoever can be *called* 'good' and 'being' insofar as it participates, through the mode of its assimilation however remotely and deficiently" in the absolute good being, who was God.[62] Similarly, natural law was natural as participatory in the same extrinsic principle. Although participation was Plato's metaphysical doctrine,[63] Aquinas did not assert participation in the Platonist forms. He posited participation in the Aristotelian principle of action. As he interpreted Aristotle's nature, the animal principle of movement was not a naturally innate form but one accepted through sense. Thus animals moved to act by natural instinct, by the form sensorially apprehended, and human animals more perfectly moved to a superior act by reason and intellect.[64] For Aquinas, the natural was not substantive but accidental, not innate but participatory. The natural law was natural not because of any intrinsic principle, any subsisting nature, but by participation in another nature, God, whose eternal law was essentially extrinsic to the rational creature.

Aquinas's argument on law involved grace, although he strictly distinguished grace from nature as only potential before its divine bestowal on human nature. He introduced grace in his questions on law as an extrinsic principle of motion parallel to law. Then he named grace a "new law." As supernaturally bestowed, grace was irreducible to nature. As he stated, "The gift of grace exceeds all created nature, since it is nothing other than some participation of the divine nature." In conclusion, "It is thus necessary that God alone deifies, communicating the society of the divine nature through some participation of likeness."[65] In sum, Aquinas posited human participation in God by similitude as defining both nature and grace. Nature participated in his eternal law; grace participated in his divine nature.

Participation was a substantial aspect of Aquinas's metaphysics,[66] and it has been related to his natural law.[67] Yet, how did he understand natural law as natural when the rational human only acquired it by participation in the eternal law? Aquinas responded that the law was natural because it was imprinted on human reason by the Creator as divine governor. That response required an explication. Since Aquinas declared all law an extrinsic principle of action, how was the natural law intrinsic to humans? As he replied, through their soul as rational. Aquinas defended his usage of "natural law" by reiterating, about the natural basis of reason and will, that the term "nature" was polysemous. His question on whether the will moved to anything naturally cited Aristotle's *Physica* on the distinction between voluntary and natural movement. It repeated Aristotle's definition of the natural as something always innate. Aquinas then took an Aristotelian line of reasoning from the senses of a word. He responded that nature had two meanings, as the intrinsic principle of mobile things, and as "any substance or any being." The latter nature meant what was consistent with a thing, thus in it. Aquinas argued that whatever was not in something per se reduced into what was in it per se, as into a principle. "And, therefore, it is necessary that, taking nature in this way, the principle in these matters that are consistent with a thing are always natural."[68] By that reasoning the natural law was natural because, as a rational principle, it was consistent with the nature of human reason, which was ordained by the good to its end. For Aquinas, because grace was also consistent with that human end, it would also be natural. That reasoning confounded his argument for grace as exceeding all nature, however. Ultimately, his dual meanings of nature collapsed. For, what was consistent with living beings, thus in them, was motion; and their intrinsic principle was also motion. For animate beings, his distinctions coincided.

Was natural law only consistent with human reason in Aquinas's substantive meaning of nature? Or was natural law also an intrinsic principle of movement in his physical meaning of nature? He certainly argued motion for the eternal law in which the natural law participated. "The reason of the divine wisdom, moving all things to the owed end, holds the reason of the law. And according to this, the eternal law is nothing other than the reason of the divine wisdom according to what is directive of all acts and motions." The participation of human reason in that eternal law defined the natural law. Therefore, there was more to Aquinas's natural law than rational consistency. There was rational movement—movement toward the designated end. Aquinas mixed the two meanings of "nature" he had distinguished. His natural law also contradicted the physical meaning of nature as a principle of intrinsic movement because he posited all law as a principle of extrinsic movement. Again, Aquinas affirmed that knowledge of the eternal law was imprinted on human reason, for it was not knowable in its essence but by its effects, by enlightenment. All laws derived from it, directing acts to ends, through moved movers to the unmoved mover who was God. Creatures were subject to the natural law not only by knowledge but also by movement. Action and passion participated in the principle of motions, subjecting even irrational creatures to the eternal law. Rational creatures participated both by some knowledge of it through its effects and by "a natural inclination to what is consonant with the eternal law."[69]

Aquinas argued that reason was based on naturally known principles, and that appetite was derived from a natural desire toward the last end. Therefore, acts toward those ends were through the natural law. That was the participation of the eternal law in the rational creature that inclined it naturally to its proper act and end. That is, natural law moved from within, intrinsically. That conclusion differed from his introduction of law according to its extrinsic principle: God moving (*movens*) by law and grace, or the devil inclining (*inclinans*) by temptation. Aquinas posed the objection to the natural law that the law ordered human acts to their end, yet the direction of human acts toward their end was not natural. Irrational animals acted toward their end solely by their natural appetites, while humans acted for their end by reason and will. The objection arose that "there wasn't any natural law for a human." Aquinas responded that the natural law was instituted in humans by the same divine providence that created the natural law of the celestial bodies, which humans reflected as "a lesser world."[70]

From the very first question he posed in *Summa theologiae*, Aquinas affirmed that a knowledge of God's existence, however laborious or erroneous, was natural to humans. That was so because humans were directed to God as their end, happiness. They needed to know that end in order to direct their thoughts and actions there. Teleological movement was fundamental, governed by the science of movement, physics. Aquinas quoted Aristotle's *Physica* that "movement is like a kind of life by nature in all existing things." All natural things moved. Movement was a norm against which Aquinas decided even divine attributes. Because God alone was immoveable, he was both changeless and eternal, thus self-sufficient and timeless. Consistent with that physics, Aquinas's first proof for the existence of God was from motion. It was certain from sensory evidence, he stated, that some things in this world were in motion. But whatever was moved was moved by another. Its movement required potentiality toward the object of movement, while the movement itself was act. To avoid an infinite regress of moved movers for enactment, Aquinas reasoned along the chain of moved objects to God as their first unmoved mover.[71] That argument excluded the possibility of self-moving things. It rejected Aristotle's principle of intrinsic movement as the definition of the natural.

Law

For Aquinas, the precepts of the natural law for the practical reason paralleled the first principles of demonstration for the speculative reason. Those precepts and those principles Aquinas declared self-evident.[72] Rational self-evidence paralleled natural self-movement; both were intrinsic and teleological. Aquinas stated the first precept of the natural law as "good ought to be done and pursued, and evil ought to be avoided."[73] Complementary to his philosophical arguments for the natural law, his introduction and definition of it relied on the *Glossa ordinaria*, the standard biblical commentary. He partially quoted Romans 2:14, "When the nations, who do not have the law, naturally do those things that belong to the law." He followed from the *Glossa*, "Although they do not have the written law, nevertheless they have the natural law, whereby anyone whoever perceives and is conscious of what is good and what is evil."[74] Aquinas argued his natural law eclectically. His other named sources were: Aristotle, *Physica*, *Metaphysica*, and *Ethica nicomachea*; Basil of Caesarea, *In Genesin*, creational homilies; John Damascene, *De fide orthodoxa*, a doctrinal summary; Augustine, *De bono coniugium*, an apologetic on marriage; Boethius, *De hebdomadis*, philosophical investigations on being; Justinian, *Digestum*, a codification of classical jurists; Gratian, *Decretum*, the basic compilation of canon law; Isidore of Seville, *Etymologiae*, an encyclopedic dictionary; Caesar, *De bello gallico*, political propaganda; and Augustine, *Confessiones*, personal epideictic rhetoric.[75]

Aquinas rephrased the first precept of the natural law without attribution from the same source that he indirectly had his psychology of the passions already treated in the same part of *Summa theologiae*.[76] That source was the compilation of a bishop in the late fourth or early fifth century, Nemesius of Emesa's *Premnon physicon sive peri physeōs anthrōpou liber*. The work was translated into medieval Latin twice. An incomplete version was *Premnon physicon* by Nicolo Alfano, a bishop of Salerno. That city was the site of the first medieval medical school,[77] and the translation was preserved and consulted principally for its medical information.[78] The complete work was translated as *De natura hominis* by Richard Burgundio of Pisa, a professor of law at Pisa and a notable translator of Aristotle and Galen. Aquinas used his translation, which misattributed the work to Gregory of Nyssa, not a church father but revered as a saint.[79]

De natura hominis was blessed by John Damascene's unacknowledged citations of it in *De fide orthodoxa*, a major patristic summary of Christian doctrine.[80] Aquinas's own theology professor Albert the Great also cited Nemesius's misattributed work.[81] It was useful as the first Christian anthropology and particularly as the first integration of Aristotelian philosophy with Christian doctrine. Its importance has been reappraised as original and philosophical for its mediation of Aristotle to Plato on the body-soul relation. Nemesius argued from their Platonist instrumentality toward their Aristotelian unity,[82] a position relevant for Aquinas's doctrine of the soul as the form of the body.[83] Aquinas used Nemesius's book for both Christian doctrine and historical reference. He consulted it in the first part of the second part of *Summa theologiae* on human acts. The topics he cited were the voluntary and involuntary, choice and counsel, appetite and reason. Aquinas also gleaned from Nemesius some opinions of ancient philosophers—Egyptian, pre-Socratic, Platonist, Stoic, and Epicurean. He also relied on Nemesius through John Damascene's unacknowledged quotations of him in *De fide orthodoxa*.[84]

Although Aquinas copied Nemesius freely, not every dependence has been recognized. For example, Aquinas's argument about whether intoxicated acts were voluntary or involuntary responded to him.[85] A major borrowing undetected by editors and commentators was Nemesius's topic of the rational nature. Aquinas rephrased it as the first precept of the natural law. Nemesius considered humans to be constituted as the "midpoint between the bounds of irrational and rational nature." Aquinas similarly designated humans as creatures intermediate between the higher angels and the lower animals.[86] Nemesius proposed in his first chapter the topic Aquinas adapted for the natural law. "The capital of the rational nature is to flee and also to avoid evils, on the contrary to go toward and choose goods (*Rationalis autem naturae capitulum est fugere quidem et avertere mala, pertransire vero et eligere bona*)."[87] Aquinas rephrased it in his question on the natural law as "The first precept of the law is that good ought to be done and pursued, and evil ought to be avoided (*Hoc est ergo primum praeceptum legis, quod bonum est faciendum et prosequendum, et malum vitandum*)."[88] He reversed the order of good and evil in conformity with his arguments that every agent acts toward a good end and that law directs toward it.[89] He revised Nemesius's election of the good to doing the good in conformity with his own argument that the will, not the reason, chose good; for, natural law concerned reason.[90] Significantly, Aquinas reformulated Nemesius's grammar from an indication to what he mistook as an absolute command.

Nemesius's term *kephalaion*, in Burgundio of Pisa's translation *capitulum*, in general meant a chief part or point. Nemesius's particular reference was to Aristotle's logic, in accord with his own modifier, *logikēs physēos*. By a principal topic Nemesius meant an Aristotelian premise for arguments. Aristotle's *Topica* distinguished reasoning as either demonstrative or dialectical. A demonstrative premise was true and primary, and from it necessary arguments could be developed. Such a premise was believable in and by itself, like the first principles of science, which needed no further inquiry. In contrast a dialectical premise afforded reasoning from generally accepted opinions.[91] Nemesius's topic of the rational nature was demonstrative. It fit Aquinas's statement that the precepts of the natural law for the practical reason paralleled the first principles of demonstration for the speculative reason. They were self-evident.[92]

Aquinas converted Nemesius's *capitulum naturae* to *praeceptum legis*. The noun *capitulum* was a common term in the *Corpus iuris civilis* for a "chapter" as a legal division. However, Aquinas's first precept of the natural law was not a legal prescription but a logical premise. Aquinas declared the first precept of the natural law as the equivalent, for the practical reason, of the first principles of science for the intellective reason.[93] According to Aristotle's logic it was demonstrative. It was a true and primary premise from which necessary arguments could be reasoned to necessary conclusions. However, the conclusions as reasoned from it were only logically necessary, not morally necessary. Aquinas's logical context differed from modern acceptance of his natural law as the "command" of the practical reason.[94] Demonstrative logic explained why Aquinas did not cast the first precept of the natural law in the imperative mood with the psalmist as "depart from evil, and do good (*diverte a malo, et fac bonum*)" (Ps. 33:15 Vulg.). Nor did Aquinas even mention the natural law in his commentary on that verse. On a similar verse, *decline a malo, et fac bonum* (Ps. 36:19 Vulg.), he interpreted its imperative mood as an "exhortation."[95]

Instead of employing the imperative mood for the first precept of the natural law, Aquinas constructed a gerundive with an indicative: *bonum est faciendum et prosequendum, et malum vitandum* ("Good ought to be done and pursued, and evil ought to be avoided").[96] A Latin gerundive was a passive and futuristic verbal adjective. It was "often used as an adjective implying obligation, necessity, or propriety (*ought* or *must*)." Combined with the verb *esse*, "to be," a gerundive formed the passive periphrastic construction. It "denotes *obligation*, *necessity*, or *propriety*."[97] Aquinas's first precept of the natural law translated literally as "good is needing to be done and pursued, and evil is needing to be avoided." He did not supply any actors or avoiders. Aquinas's precept was not direct speech as an imperative command that intended movement to action. Its denotation only indicated an obligation.

An interpretation of Aquinas's natural law as not a moral command has noted his use of a gerundive, rather than an imperative, for its first precept. It argued that, in contrast to an imperative, a precept expressed in the gerundive form "merely offers rational direction without promoting the execution of the work to which reason directs." Then it altered its Latin grammar. It asserted that, although Aquinas's gerundive was "not primarily imperative force," nevertheless it was "really prescriptive" and "not merely a theoretical statement." That was so because "precepts do not inform us of requirements; they express requirements as directions for action." Then it erred in grammar. "Of course, so far as grammar alone is concerned, the gerundive form can be employed to express an imperative." It claimed that the difference between the imperative "do good and avoid evil" and the gerundive "good is to be done and pursued and evil is to be avoided" was "the omission of *pursuit* from the one, the inclusion of it in the other." By "pursuit," Aquinas thus asserted the relation of the first precept of the natural law to final causality, a new achievement.[98] However, Nemesius's topic of the rational nature had before Aquinas employed continuous actions of fleeing evils and going toward goods.[99] That connotation of ends was consistent with the Aristotelian teleology in Nemesius's compilation of a Christian anthropology.

Aquinas himself had in the same part of his *Summa theologiae* addressed the grammar of command. Deliberating whether to command (*imperare*) was an act of the reason or of the will, he responded on the supposed authority of Gregory of Nyssa (actually Nemesius) about Aristotle's psychology. "The appetite obeys reason, therefore it belongs to reason to command (*imperare*)." Aquinas acknowledged that the rational command presupposed an act of the will.[100] But he argued, "To command (*imperare*) is indeed essentially the act of reason, for the one commanding disposes (*ordinat*) him whom he orders toward doing something, by announcing (*intimando*) or officially announcing (*denuntiando*); for thus to dispose by means of some announcement (*intimatio*) belongs to reason." He argued further. "But reason can announce (*intimare*) or officially announce (*denuntiare*) something in two ways. By one mode, absolutely (*absolute*), which announcement (*intimatio*) is expressed through a verb in the indicative mode, just as someone says to someone 'this ought to be done by you' (*hoc est tibi faciendum*)."[101] Two questions later Aquinas distinguished a necessary saying *absolute*, or unconditionally, from a necessary saying *ex suppositione*, or conditionally. "Something is judged necessary absolutely from the condition of the terms, namely because the predicate is in the definition of the subject, as it is necessary that man is an animal; or because the subject belongs to the reasoning of the predicate, as it is necessary that a number is odd or even."[102]

Aquinas compared commands. "Sometimes, moreover, reason announces something to someone by moving (*movendo*) him to it, and such an announcement is expressed through a verb in the imperative mode, for example, 'do this' (*fac hoc*)."[103] Aquinas distinguished rational commands thus: the gerundive form in the indicative mood announced a command absolutely, while the imperative mood announced it by some motion. He did not discuss there the use of the jussive subjunctive to command. However, in the third part of his *Summa theologiae* he mistook a famous biblical example (Genesis 1:3) of the jussive subjunctive, *fiat lux* "let there be light," for the imperative mood. He maintained that the verse expressed the Creator's "efficacy through a command" by "a verb in the imperative mode."[104] However, *fiat* "let there be" was in the subjunctive mood. Even in Aquinas's prior argument on rational command his grammatical argument for the gerundive form was incorrect. Yet, he referred to that argument in his question on the law, for "it belongs to reason to ordain to the end, which is the first principle in acting, according to Aristotle." That referenced his *Physica* on the types of necessity.[105]

Aquinas stated that the precepts of the natural law for the practical reason paralleled the principles of demonstration for the speculative reason. Both were self-evident.[106] He formulated the first precept in the passive periphrastic construction, based on his prior assertion that its grammar, in the indicative mood, was an absolute, or logically necessary, command.[107] In Latin grammar the indicative mood was used for a statement or question; the imperative mood, for a command. Aquinas's gerundive *intimando* occurred in his entire works only in the one question about rational command. His infinitive *intimare*[108] occurred a dozen times, mostly in quotations, about the angelic or human communication of a concept. None of his examples of *intimo* issued commands.[109] They did not support his claim that *intimo* commanded. They cohered, however, with his distinction in a later question on law between an utterance and a law. "Just as an utterance (*enuntiatio*) is a saying (*dictamen*) of the reason through the mode of uttering (*enuntiandi*); so thus a law (*lex*) through the mode of giving rules (*praecipiendi*)." He argued that "just as in the demonstrative sciences reason induces assent to a conclusion through certain principles, thus also it induces assent to a rule of the law through something else."[110] Those particular gerundives, *enuntiandi* but especially *praecipiendi*, to adduce assent may have prompted Aquinas to cast his first principle of the natural law in the passive periphrastic form.

However, the passive periphrastic construction of Aquinas's first precept of the natural law was not grammatically a command to do and pursue good, and to avoid evil. It was a denotation, an indication: *bonum est faciendum et prosequendum, et malum vitandum* ("Good is needing to be done and pursued, and evil to be avoided"). What sort of necessity, obligation, or propriety did the passive periphrastic construction indicate according to Latin authors? The most famous example was *Carthago delenda est* ("Carthage ought to be destroyed"). It characterized the rhetorical, not philosophical or legal, nature of the usage of the passive periphrastic construction. *Carthago delenda est* was the slogan of the Roman senator Cato the Elder, whom Aquinas respected as an exemplar of piety, virtue, and wisdom. He even bowed to Cato in explicating the Apostles' Creed.[111] Aquinas could have known of Cato's legendary political hostility to Carthage from various sources, such as Augustine's *De civitate Dei*, which he cited frequently. Cato's reputation, as Livy reported it by a gerundive of purpose, was as "a man quick-tongued in the senate for vituperating (*ad vituperandum*)." In that role as the snappy patriotic opposition Cato ended his every senatorial vote with the rider, "And in my opinion,

'Carthage ought to be destroyed.'" His demand was first cast in the passive periphrastic construction by Pliny the Elder's *Historia naturalis* in its disquistion on figs. It told how Cato brandished a fresh fig before the senators, warning them that Carthage, where it had been picked, was but a three-day sailing from Rome. Although some senators thought Cato's animosity irrelevant, and Scipio confronted it with "Carthage ought to stand," the Romans eventually waged the third Punic war and razed Carthage. Pliny marveled that the clever display of a fruit toppled a nation. Plutarch's *Lives* retold the story, detailing how Cato shook the fig from the folds of his toga. "And on one issue he was even more savage, namely in adding to his vote on any question whatsoever these words: 'In my opinion, Carthage ought to be destroyed.'" The anecdote was sure to enliven a grammar lesson on the passive periphrastic construction, and the phrase was very likely among Roman school exercises. Its vigorous description demonstrated the passionate political context for Cato's epigram. The phrase was dramatically, near sensationally, declamatory. In modern English it is "a standing description of a 'bitter-ender' and the ready formula of celebrated chauvinists and jingoes."[112] Cato's consistent verb was *censeo*, literally "to tax," whence a census. Tropologically it meant "to be of the opinion, to propose, to vote, to move." At most Cato was proposing a senatorial vote, a republican exercise foreign to any medieval state in which Aquinas taught. It was not legal command that "Carthage ought to be destroyed." It was only Cato's individual political opinion stubbornly and loudly insisted on until the Roman senate conceded to act upon it. As Livy characterized Cato's soulful passion, he was "afire with consuming hatred of Carthage and worried for the security of his grandchildren when he used to shout at every senate 'Carthage ought to be destroyed.'" Laws did not clamor; they decreed. Cato's hollering was not even in the deliberative genre of rhetoric designated for civic discourse in the senate. It belonged, as a judgment of blame, to the epideictic genre. Cato was renowned as a moral man. But Cato's passive periphrastic construction, "Carthage must be destroyed," however formed in moral judgment, in grammatical construction was not a law. Aquinas's substitution of general moral evil—*malum*—for a particular political enemy—*Carthago*—did not aggrandize or alter the function or force of the passive periphrastic construction. It did not render Aquinas's first precept of the natural law a command. Aquinas's own use of the gerundive for *delere* mostly concerned the remission of sins (*ad delendum peccata*).

As for *faciendum* in his first precept of the natural law, he wrote that gerundive hundreds of times, one hundred seventeen in *Summa theologiae* alone. It usually had *aliud* "something" as its unspecified object, but it occurred in the context of doing good and avoiding evil. Before Aquinas's statement of the first precept of the natural law, however, he constructed the passive periphrastic *faciendum + esse* only seven times. The initial case concerned the triple roles of conscience: to recognize and witness; to judge and to stimulate or bind; and to judge and to excuse or accuse, or torment. In its second role conscience judged that "something ought to be done or not to be done (*iudicamus aliquid esse faciendum vel non faciendum*)."[113] His next passive periphrastic construction was posed as an objection that command is not an act of the reason. "For he who judges that something ought to be done (*iudicat aliquid esse faciendum*) does not perform it right away." That case occurred in the article with his mistaken grammar for command.[114] His next case was hypothetical as cast in the subjunctive mood. "If reason or conscience should say (*si … dicat*) that something out to be done (*aliquid esse faciendum*) because it might be good generically, there is no error. Similarly if it should say (*si dicat*) that something ought not to be done (*aliquid non esse faciendum*) because it is generically evil, good deeds are ruled (*praecipuntur*) from the same argument by which evil deeds are prohibited."[115] There were two further cases that something "should not be done (*non esse faciendum*)."[116] Then in his questions on the law Aquinas constructed a significant gerundive of purpose. "It belongs to the law properly to oblige toward doing or not doing something (*obligare ad aliquid faciendum vel non faciendum*)." The definition of "obligation" in Justinian's legal *Digest* was "The substance of obligations consists in binding us either to giving, or doing (*faciendum*), or warranting something." Justinian's legal obligation of doing (*obligatio … ad faciendum*) differed from Aquinas's precept of the natural law that something "ought to be done" (*faciendum est*). Law imposed an obligation; Aquinas's precept only indicated it.[117]

Then followed Aquinas's passive periphrastic construction of the first precept of the natural law, *bonum est faciendum et prosequendum, et malum vitandum* "good ought to be done and pursued, and evil ought to be avoided."[118] However, his further passive periphrastic construction redefined law. Reiterating that "the natural law was the law imparted to humans," Aquinas introduced the superaddition of the gift of grace with double gerundives. Of grace, "By this means a new law is implanted in a human, not only indicating what might be done (*indicans quid sit facien-*

dum) but even assisting toward its implementation (*ad implendum*)."[119] What sort of law indicated and assisted? Indication was from the *index* finger, in medieval art the primary gesture of the teacher of a discipline and also of the personifications of grammar and wisdom.[120] The index finger was called *demonstratorius* because in medieval etymology it signaled.[121] Indication corresponded to Aristotle's type of logical reasoning called demonstrative, or true and necessary. But Aquinas's first precept of the natural law, as his self-evident premise of the practical reason, did not deduce conclusions that were legally binding. He did argue that particular conclusions could be derived from the common principles of the natural law, just as a house builder worked from an architectural prototype.[122] However, conclusions deduced from his first precept of the natural law were only necessarily logical.

Aquinas developed his arguments on law by supposing that the divine lawgiver disposed his various creatures with various inclinations. From their "inclinations (*inclinationes*)" he concluded their "law (*lex*)." He posited that "somehow there is the law of the dog that it ought to rage (*furibundum esse est quodammodo lex canis*)." Again, the grammar, *furibundum esse*, was a passive periphrastic construction, like *faciendum esse* in his first precept of the natural human law.[123] However, Aquinas's canine "law" was only an example from Galenic humoral theory in medicine, which typified physiological constitutions, not natural laws. As Aquinas's teacher Albert the Great wrote in *De animalibus*, certain animals were ferocious because of their hot and dry temperaments. Others, like the sheep that his student Aquinas opposed to the dog, were meek because of their cold and wet temperaments.[124] From the canine "law" Aquinas concluded, "Therefore, there is the human law, which is allotted by divine arrangement according to his proper condition, that he might operate according to reason."[125]

Aquinas had adapted Nemesius's topic of the rational nature for deriving necessary arguments for the behavior of humans as creatures "midpoint between the bounds of irrational and rational nature." His reliance on that anthropology began appropriately in his questions on humans, specifically about the error of the ancients on the soul as a kind of body. But Aquinas borrowed from Nemesius's book fully, from its first to its forty-fourth chapter. He accepted from Nemesius's first chapter the premise of a human being as "a lesser world" between the bounds of rational and irrational beings. Then he applied it to a human being as between the bounds of spiritual and corporeal substances. Microcosmic man was Nemesius's foundation for his topic of the rational nature, which

Aquinas adapted for the first precept of the natural law. Aquinas anticipated that precept in reflecting on Nemesius's division of the sensible appetite into concupiscible and irascible passions. Aquinas declared that in natural corruptible things there was a necessity for "an inclination to pursuing the agreeable and fleeing the harmful (*inclinationem ad consequendum conventientia et refugiendum nociva*)." His grammar was constructed by gerundives, *consequendum* ... *refudiendum*. They did not denote obligation but indicated the weaker movement of "an inclination." That inclination influenced Aquinas's first precept of the natural law, "Good ought to be done and pursued, and evil ought to be avoided." Adherence to it followed natural inclination. "Because truly the good has the reason of the end, moreover evil the reason of the opposite, thence consequently it ought to be pursued by work and the contraries of these so that evils are also avoided." Aquinas detailed triple human inclinations: to the universe, to other animals, and to their unique rational nature. All those inclinations tended to the ends of the preservation, perpetuation, and promotion of life.[126]

Contexts

The historical contexts, theoretical and practical, of Aquinas's *Summa theologiae* informed its arguments for the law of the heart as natural reason. He composed it not as a freelance thinker but as a friar vowed in obedience to teach the ecclesiastical tradition to beginners in theology within his religious Order.[127] Toward executing that responsibility he employed the disciplines of theology and philosophy as distinct yet complementary. Aquinas's eleven Aristotelian commentaries and an index, written simultaneously with that *Summa*, demonstrated his dual commitments. His intended coordination of reason and faith occasioned some issues, however. For his development and articulation of the natural law, Aquinas depended on Aristotle's philosophy for the physics of movement and for the psychology of reason. But Aquinas's efforts to give reason its due were confronted by the contradictions between the premises of Aristotle's uncreated universe and his own created universe, and between the motive principles of an intrinsic or an extrinsic nature. Those contradictions extended to Aquinas's definition of natural law and to his predication of nature and of law as intrinsic and extrinsic. As his resolution he posited a metaphysics of rational participation in God's mind. His adherence to the Aristotelian primacy of reason, however, occasioned the issue of how reason morally commanded an act rather than logically concluded to it. Aquinas's source for his first precept of the natural law was Nemesius's topic of the rational nature, which he adapted. As a premise for Aristotelian logical demonstration, its binding on reason was only logical. Aquinas's metaphysics of human participation in divine reason radically altered Aristotle's demonstrative logic. For Aquinas, a rational conclusion about good or evil became a command because human reason participated in the divine mind.

Aquinas's first precept of the natural law for the practical reason, as self-evident, did not logically make it lawfully binding, however. None of his paralleled examples of self-evident principles for the speculative reason, beginning with non-contradiction, commanded.[128] Although Aquinas frequently wrote the passive periphrastic construction, that construction only grammatically implied or indicated necessity or obligation. It did not command. From its premise a conclusion could be reasoned, but that conclusion was only logically necessary, not morally binding. The question remained how Aquinas's natural law as an indication that was self-evident and a premise for demonstrative reasoning related to any obligation to act upon it. Aquinas believed that human reason derived from its creation in the image of the divine mind. It was thus fulfilled in acts consistent with its inclination to God as its origin and end. He posited that reason commanded the observance of the natural law, whereas reason only logically concluded. He did not discuss whether everything reasonable about good was obligatory for doing and pursuing it or everything reasonable about evil was obligatory for avoiding it.

Aquinas's law of the heart also evidenced some cultural issues of scholarship. His childhood education in the trivium was at the venerable but then decadent Benedictine abbey of Monte Cassino, where his father's brother was the abbot. When the emperor Frederick II expelled most of the monks, the new abbot advised his parents to consent to transfer then adolescent Thomas to the emperor's charter university at Naples, which he founded to train his bureaucrats in law. There Aquinas was further educated in grammatical and logical subjects.[129] His eventual argument in *Summa theologiae* for the first precept of the natural law evidenced some deficiency in Latin grammar. His account of verbal moods—indicative, imperative, subjunctive—and their function to command was neither classical nor medieval usage. Aquinas's monastic education in rhetoric was also at issue. His use of quotations from Terence's *Eunuchus*, a medieval school text, did not understand the plot of that play. Aquinas abstracted logically the heart in the rhetoric of Augustine's *Confessions* to argue his own theory of its natural law. Aquinas was further culturally disadvantaged by uncritical manuscripts and mistranslated verses of Scripture to which as a theologian he appealed as authoritative. That reality was commonplace for medieval thinkers, who lacked the historical method and philological science for reading the Bible as it was originally transmitted. Aquinas personally lacked the ability to read the Bible in the available Hebrew and Greek manuscripts. Thus he unknowingly conscripted as proofs for natural law biblical verses that were mistranslations. Not only were some biblical texts unreliable but also some patristic interpreters were, including Augustine, whom Aquinas cited most often but who also lacked the historical and linguistic skills for a literal biblical knowledge. However, Aquinas's exegetical practice reflected his ecclesiastical reality. The fellow Dominican friars who were in his charge to teach were required in their ministry to use the Bible as translated in their lectionaries.

The practicality of his *Summa theologiae* accounted further for unclarity about his natural law of the heart. That reference work was not a speculative monolith but his accommodated pedagogy. Aquinas initiated the project in 1265 at the Dominican house of studies in Rome, a novel personal appointment in his religious Order of Preachers. His office was to teach not philosophers and theologians but friars, specifically the "juniors." Those were the great majority of the Dominican friars, at any age, who lacked the benefit of a university education. Aquinas's innovation was to augment their available manuals for preachers and confessors by situating their pastoral ministry in a doctrinal context.[130] The preface to his *Summa theologiae* intended "to treat the matters that pertain to the Christian religion in a manner conforming to the education of beginners." Aquinas meant "to be concise and clear so far as the subject allows."[131]

Its question on the natural law belonged to his second term at Paris as regent master of the Dominican house of studies there. Labor on the entire second part of his *Summa theologiae* was intense, as he dictated to secretaries its 303 questions in eighteen months. That regency from 1268 to 1272 also produced such diverse works as commentaries on the gospels of Matthew and John, the disputed question *De malo*, and six quodlibets, or responses to impromptu academic topics.[132] There is no evidence that Aquinas ever actually taught his *Summa theologiae*.[133] Exactly what its availability in manuscript was to his students and what use they made of it are unknown. Its fortune was unlike that of his biblical lectures as an academic *magister in sacra pagina* for which some notes by students and scribes are extant. After Aquinas's death his brethren finished that *Summa* with cut-and-paste from his writing on Peter Lombard's *Sententiae*. The Dominicans then circulated the three parts of *Summa theologiae* independently. For their ministry, they were keenest for the practical second part of its second part about virtues, gifts, and the religious life. Its circulation almost doubled that of the theoretical first part of the second part with the question on the natural law. The part with the natural law accounted for only 20 percent of all manuscripts.[134] The Dominicans, who were founded in 1215 with the office of preaching, were in the next decade papally commissioned to educate confessors and to administer the sacrament of penance. The natural law did not require Aquinas's great attention because it was not essential to their ministry. The friars for whose consultation he composed *Summa theologiae* would be hearing the sacramental confessions of Catholics who broke the divine commandments by willful sins. They would not be hearing any confessions of anyone who broke the natural law by irrational conclusions.

On a question about penance Aquinas famously quit his *Summa theologiae* in December 1273 and never wrote another word.[135] Its question on the permanence of the natural law in the heart had simply stated his minor premise that "the law written in human hearts is the natural law."[136] However, in 1272–1273 Aquinas's fraternal lecture on the Epistle to the Romans[137] identified its verses as his source for the law of the heart as the natural law. As he cited more fully this time, "For when the nations, who do not have the law, naturally do those things that are of the law, in such a manner not having the law, they are a law to themselves who manifest the work of the law written in their hearts" (Romans 2:14–15). Aquinas's lecture interpreted its verse "they naturally do the things that are of the law" as "the mandates of the law, clearly in regard to the moral precepts, which are from the prescription of natural reason (*de dictamine rationis naturalis*)."[138] In Aquinas's lecture those who lacked the biblical commandments of divine revelation could and did keep them as the moral precepts of natural reason. His interpretation was consistent with his belief that God commanded and reason also commanded. Such was Aquinas's law of the heart. But ultimately he revised it. The very first commentary on Romans, which was Origen's, determined that "in hearts" should not be understood as "in the bodily member" because the physical heart could not be a receptacle of prudence or memory. Rather, "heart" meant the "rational power of the soul for perceiving."[139] However, in Aristotle's natural philosophy and its medieval tradition the seat of the soul was indeed the physical heart, in distinction to the brain. In 1273 before he quit, Aquinas wrote a brief treatise *De motu cordis* (On the movement of the heart), which rejected the Aristotelian soul resident in the heart. Aquinas asserted his own doctrine of the soul as the form of the body but not as resident in the heart.[140] The natural law of the rational soul was no longer the law of the heart.

Notes

1. Thomas Aquinas, *Summa theologiae* II-I, q. 91 a. 2; q. 94. See also q. 95 a. 2. For the Leonine edition, a critical revision of which is in progress, see http://www.corpusthomisticum.org/.
2. Aquinas, *Summa theologiae* II-I, q. 91 a. 2. "... *dicit Glossa, 'etsi non habent legem scriptam, habent tamen legem naturalem, qua quilibet intelligit et sibi conscious est quid sit bonum et quid malum*." Cf. *Biblia latina cum glossa ordinaria*, facsimile rpt. of Strassburg: Adolph Rusch, 1480–81, 4 vols. (Turnholt: Brepols, 1992) ad loc. "*si non habeat scriptam legem habet naturalem, quia intelligit et sibi conscius est quid sit bonum et quid sit malum*." Anselm of Laon was its editor, according to multiple sources, for the Pauline epistles. E. Ann Matter, "The Church Fathers and the *Glossa ordinaria*," in *The Reception of the Church Fathers in the West: From the Carolingians to the Maurists*, ed. Irena Backus, 2 vols. (Leiden: Brill, 1997), 1:107. In general see Lesley Smith, *The "Glossa ordinaria": The Making of a Medieval Bible Commentary* (Leiden: Brill, 2009).
3. *Glossa ordinaria* ad. loc. The gloss attributed to Origen is apparently an editorial summation of Rufinus's Latin abridgement of Origen's commentary on Romans. For the law written in their hearts as "what they can sense naturally," see Origen, *Commentarii in Epistulam ad Romanos* 2.9.3, in *Römerbriefkommentar des Origenes*, ed. Caroline P. Hammond Bammel, 3 vols. (Freiburg: Herder, 1990–97); and rpt., 4 vols. (Paris: Cerf, 2009–12). For the diffusion of Origen's works in Latin copies, see Henri de Lubac, *Exegese médiévale: Les quatres senses de l'Écriture*, 2 vols. in 4 (Paris: Aubier, 1959–64). Cf. Peter Lombard's theological alteration of Origen in the *Glossa* to "by natural reason illuminated through grace." Peter Lombard, *Collectanea in omnes d. Pauli apostoli epistolas*, in *Patrologia cursus completus series latina*, ed. J.-P. Migne, 221 vols. (Paris, 1800–75), 191:1345–46.
4. *Glossa ordinaria* ad loc. Lombard, *Ad Romanos* ad loc. identifying "heart" with affect, where faith operates through love, as Ambrose [Ambrosiaster] although it is not there. See Ambrosiaster, *Commentaria*

in XII epistolas beati Pauli, in *Patrologia latina*, 17:71, but *per rationem naturae* and *natura duce.*

5. Augustine, *Confessions libri tredecim* 2.4.9, ed. Lucas Verheijen (Turnout: Brepols, 1981).
6. *Summa theologiae* II-I, q. 94 a. 6.
7. Marjorie O'Rourke Boyle, "Augustine's Law of the Heart: Thieves' Honor," in this volume, pp. 35–66.
8. Marjorie O'Rourke Boyle, "The Prudential Augustine: The Virtuous Structure and Sense of His *Confessions*," *Recherches augustiniennes* 22 (1987): 129–50; see also idem, "A Likely Story: The Autobiographical as Epideictic," *Journal of the American Academy of Religion* 57 (1989): 23–51.
9. See Theodore C. Burgess, *Epideictic Literature* (Chicago: University of Chicago Press, 1902).
10. Augustine, *Retractationum libri duo* 2.6.1, ed. A. Mutzenbecher (Turnhout: Brepols, 1999).
11. *Summa theologiae* proem.
12. Cicero, *De inventione* 1.7.9.
13. Augustine, *Retractationes* 2.5.1; Boyle, "Prudential Augustine."
14. Cicero, *De inventione* 1.5.7; 2.59.177; 1.24.34–36; *Topica* 23.89. Burgess, *Epideictic Literature*, pp. 119–26.
15. Cicero, *De oratore* 2.5.20.
16. Boethius, *De topicis differentiis* 4, in *Patrologia latina*, 64:1173–1216; *Boethius's "De topicis differentiis,"* trans. Eleonore Stump (Ithaca, N.Y.: Cornell University Press, 1978), pp. 79, 80.
17. *Summa theologiae* II-I, q. 94 a. 2.
18. Ibid. a. 6.
19. Augustine, *Confessions* 2.3.8, 2.4.9.
20. Boyle, "Augustine's Law of the Heart."
21. Augustine, *De civitate Dei* 4.4., ed. Bernhard Dombart and Alphonse Kalb, 2 vols. (Turnhout: Brepols, 1995). Cited by Brent D. Shaw, "Bandits in the Roman Empire," *Past and Present* 105 (1984): 3; Nicholas K. Rauh, *Merchants, Sailors, and Pirates in the Roman World* (Stroud: Tempus, 2003), p. 195. By this date Augustine's knowledge of their habits may owe also to Apuleius, *Metamorphoses* 3.27–4.22, with sharing the loot at 3.28 and disguises at 4.14–15, 7.8. For their vocabulary, see Werner Riess, *Apuleius und die Räuber: Ein Beitrag zur historischen Kriminalitätsforschung* (Stuttgart: Franz Steiner, 2001), pp. 32–44. See also Vincent Hunink, "'Apuleius, qui nobis Afris Aer est notior': Augustine's Polemic against Apuleius in *De civitate Dei*," *Scholia: Studies in Classical Antiquity* 12 (2003): 82–95. For other fictional authors on the pirate share, see Philip de Souza, *Piracy in the Graeco-*

Roman World (Cambridge: Cambridge University Press, 1999), p. 216. For piracy in Homer, whom Augustine read in school, see pp. 17–21.

22. *Summa theologiae* II-II, q. 66 a. 8 ad 3.
23. Boyle, "Augustine's Law of the Heart."
24. Augustine, *De civitate Dei* 4.1.
25. Boyle, "Augustine's Law of the Heart."
26. His grammar schooling was at the Benedictine abbey of Monte Cassino. Jean-Pierre Torrell, *Saint Thomas Aquinas*, vol. 1: *The Person and His Work*, rev. ed., trans. Robert Royal (Washington, D.C.: Catholic University of America Press, 2005), p. 4. For his first biography, see Guillaume de Tocco, *Ystoria sancti Thome de Aquino, de Guillaume de Tocco (1323)*, ed. Claire le Brun-Gouanvic (Toronto: Pontifical Institute for Mediaeval Studies, 1996), pp. 100–2. For Terence as a major author in medieval Italian schooling, see Jessica Levenstein, "Terence," in *Medieval Italy: An Encyclopedia*, ed. Christopher Kleinhenz, 2 vols. (New York: Routledge, 2004), 2:1072. For context, see Francis Newton, *The Scriptorium and Library at Monte Cassino, 1058–1105* (Cambridge: Cambridge University Press, 1999).
27. Aristotle, *Poetica* 4 1449a.
28. *Summa theologiae* II-II, q. 53, a. 6 ad 2.
29. Terence, *Eunuchus* lines 46, 57–63, in *Terence*, ed. and trans. John Barsby, 2 vols. (Cambridge, Mass.: Harvard University Press, 2001), 1:321.
30. *Summa theologiae* II-II, q. 153 a. 5.
31. Terence, *Eunuchus* lines 64–68, 188; trans., pp. 321, 323.
32. Boyle, "Augustine's Law of the Heart."
33. Aquinas, *De malo*, q. 15 a. 4, citing Terence, *Eunuchus* lines 57–58, 67. This dates perhaps slightly earlier to circa Paris 1270. Giles Emery, "Brief Catalogue of the Works of Saint Thomas Aquinas," in Torrell, *Aquinas*, p. 336.
34. *Summa theologiae* I-II, q. 73 a. 5; q. 75 a. 4.
35. Ibid. I-II, q. 42 a. 5; II-II, q. 162 a. 1 (supplement); I-II, q. 46 a. 2; II-II, q. 106 a. 2; I-II, q. 42 a. 5, compared with I-II, q. 45 a. 1. His citations are from Augustine, *Confessions* 2.6.13, 2.7.15.
36. *Summa theologiae* II-I, q. 90.
37. Ibid. qq. 90–108.
38. Ibid. q. 90 proem; q. 90 a. 1; I q. 82 a. 3; II-I q. 17 a. 1.
39. Ibid. qq. 90–108.
40. Ibid. q. 91 aa. 1, 2, 6.
41. Tony Burns, *Aristotle and Natural Law* (London: Continuum, 2011); Owen Anderson, *The Natural Moral Law: The Good after Modernity* (Cambridge: Cambridge University Press, 2012), pp. 46–64.

42. Jean-Pierre Torrell, *Aquinas's "Summa": Background, Structure, and Reception*, trans. Benedict M. Gavin (Washington, D.C.: Catholic University of America Press, 2005), pp. 14–15; Torrell, *Aquinas*, pp. 328–29; Emery, "Brief Catalogue," p. 342.
43. Emery, "Brief Catalogue," pp. 343–44.
44. *Summa theologiae* II-I q. 94 a. 4.
45. Ibid. I q. 18 a. 1; II-I q. 17 a. 9; I q. 115 a. 2.
46. n. 37.
47. *Summa theologiae* II-I q. 91 a. 2.
48. Dated to 1273 by Emery, "Brief Chronology," p. 341.
49. Aquinas, *In psalmos Davidis expositio* ad loc.
50. *Summa theologiae* II-I q. 91 a. 2.
51. Ibid. I q. 1 a. 1.
52. See Raphael Loewe, "The Medieval History of the Latin Vulgate," in *The West from the Fathers to the Reformation*," vol. 2 of *The Cambridge History of the Bible* (Cambridge: Cambridge University Press, 1975–76), pp. 145–48, 149.
53. The Hebrew *textus receptus* for the Authorized Version (King James Bible) was the rabbinic bible Mikraot Gedolot in its *editio princeps* (Venice: Daniel Bomberg, 1524–25).
54. For Augustine's controversy with Jerome, see Timothy M. Law, *When God Spoke Greek: The Septuagint and the Making of the Christian Bible* (New York: Oxford University Press, 2013), pp. 161–66.
55. *Vetus latina database* (Brepolis online) reporting *exalta* Codex Casin. (Amelli, 1912) and *leva* HIER.psalt.iuxta.Hebr. (de Lagarde, 1874). Based on the records of the Vetus Latina Institute in Beuron but online database available by subscription only.
56. Augustine, *Enarrationes in psalmos*, ed. Eligius Dekkers and Iohannes Fraipont, 3 vols. (Turnhout: Brepols, 1956) ad loc.
57. See Françoise Dumas, "Monnayage et monayeurs," in *Artistes, artisans et production artistique au moyen âge*," ed. Xavier Barral I Altet, 3 vols. (Paris: Picard, 1986–90), 1:483.
58. *Summa theologiae* II-I proem.
59. Ibid. I q. 93 a. 6 ad 1, a. 4.
60. Ibid. q. 92. Aquinas, *In 4 Sententiae* 15.2.5; *Summa theologiae* II-II q. 32 a. 8; *In 4 Sententiae* 32.1.2, 36.1; *Super 1 ad Corinthianos* 3.3.
61. *Summa theologiae* I q. 105 a. 2; II-I q. 6 a. 1.
62. Ibid. I q. 79 a. 3 (italics mine).
63. See Rosa Padellaro de Angelis, *L'influenza del pensiero neoplatonico sulla metafisico di s. Tommaso d'Aquino* (Rome: Abete, 1981); Pierre Faucon, *Aspects néoplatoniciens de la doctrine de saint Thomas d'Aquin* (Paris: H. Champion, 1975).

64. *Summa theologiae* I q. 18 a. 3.
65. Ibid. II-I, q. 112 a. 1.
66. John Rziha, *Perfecting Human Actions: St. Thomas Aquinas on Human Participation in Eternal Law* (Washington, D.C.: Catholic University of America Press, 2009).
67. Martin Rhonheimer, *Natural Law and Practical Reason: A Thomist View of Moral Autonomy*, trans. Gerald Malsbary (New York: Fordham University Press, 2000), pp. 241–51; Fulvio Di Blasi, *God and the Natural Law: A Rereading of Thomas Aquinas* (South Bend, Ind.: Notre Dame University Press, 2006), pp. 121–24.
68. *Summa theologiae* II-I q. 91 ad 2; q. 10 a. 1.
69. Ibid. q. 93 aa. 1, 2, 3, 6.
70. Ibid. q. 91 a. 2; q. 90; q. 91 a. 2. See also Aquinas, *De motu cordis*, ed. Fratres Praedicatores, in *Opera omnia iussu Leonis XIII P.M. edita* (Rome, 1976), 43:91–130.
71. *Summa theologiae* I q. 1 a. 1, q. 2 a. 1. I q. 18 a. l. I q. 9 aa. 1, 2; q. 10 a. 1. I q. 2 a. 3; q. 19 a. 1; q. 75 a. 1.
72. Ibid. II-I q. 94 a. 2; see also q. 91 a. 3. For self-evidence, see John Finnis, *Natural Law and Natural Rights*, 2nd ed. (Oxford: Oxford University Press, 2011), p. 34; Germain G. Grisez, "The First Principle of Practical Reason: A Commentary on the *Summa theologiae*, 1–2, Question 94, Article 2, *Natural Law Forum* 10 (1965): 172–75.
73. Aquinas, *Summa theologiae* II-I q. 94 a. 2.
74. Ibid. q. 91 a. 2.
75. Ibid. q. 94.
76. Ibid. q. 35.
77. Nemesius, *Premnon physicon sive Peri physeōs anthrōpou liber*, trans. Nicolo Alfano, ed. Carl Burkhard, rev. Friedrich Lammert (Leipzig: B. G. Teubner, 1917).
78. William Telfer, trans., *Cyril of Jerusalem and Nemesius of Emesa* (London: SCM, 1955), pp. 217–18, 206–8, 275, 370; R. W. Sharples and Philip van der Eijk, trans., introduction to Nemesius, *On the Nature of Man* (Liverpool: Liverpool University Press, 2008), pp. 11–14, 20–21, 23–25; Eiliv Skard, "Nemesiosstudien," *Symbolae osloensis* 17 (1937): 9–25; 18 (1938): 31–41; 19 (1939): 46–56; Friedrich Lammert, "Hellenistische Medizin bei Ptolemaios und Nemesios: Ein Beitrag zur Geschichte der christlichen Anthropologie," *Philologus: Zeitschrift für klassische Altertum* 94 (1940): 125–41; Alberto Siclari, *L'antropologia di Nemesio di Emesa* (Padua: Garangola, 1974), pp. 149–80.
79. Gérard Verbeke and J. R. Moncho, eds., Némésius D'Emese, *De natura hominis: Traduction de Burgundio de Pisa* (Leiden: E. J. Brill, 1975), pp. lxxxvi–xcii.

80. See Emil Dobler, *Zwei Syrische Quellen der 'Theologischen Summa' des Thomas von Aquin, Nemesios von Emesa und Johannes von Damaskus: Ihr Einfluss auf die anthropologischen Grundlangen der Moraltheologie (S. Th. I-II, qq. 6–17, 22–48)* (Freiburg: Universitäts, 2000); idem, *Falsche Väterzitate bei Thomas von Aquin: Gregorius, Bischof von Nyssa oder Nemesius, Bischof von Emesa? Untersuchungen uber die Authentizität der Zitate Gregors von Nyssa in der gesamten Werken des Thoms von Aquin* (Freiburg: Universitäts, 2001); idem, *Indireckte Nemesiuszitate bei Thomas von Aquin: Johannes von Damaskus als Vermittler von Nemesiustexten* (Freiburg: Universitäts, 2002).
81. E.g., Albertus Magnus, *De bono* 3, q. 5, a. 2 in *Opera omnia*, ed. Heinrich Kühle et al. (Monasterium Westfalorum, 1951–), 28:201. See also Moreno Morani, *La tradizione manoscritta del "De natura hominis" di Nemesio* (Milan: Vita e pensiero, 1981).
82. Beatrice Motta, *La mediazione estrema: L'antropologia di Nemesio di Emeso fra platonismo e artistotelismo* (Padua: Poligrafo, 2004).
83. See *Summa theologiae* I, q. 75–76.
84. Note 80.
85. *Summa theologiae* II-I q. 7 a. 4; cf. Nemesius, *De natura hominis* 29, 30, trans. Burgundio of Pisa, ed. Verbeke and Moncho, pp. 119–24.
86. See Gerard Verbeke, "Man as a 'Frontier' according to Aquinas," in *Aquinas and Problems of His Time*, ed. idem and D. Verheist (Louvain: Leuven University Press, 1970), pp. 195–223.
87. Nemesius of Emesa, *De natura hominis* 1, trans. Burgundio, pp. 9–10. The Greek is *tēs de logikēs physeōs to kephalaion esti pheugein men kai apostrephesthai ta kala*. Nemesius, *Peri physeōs anthrōpou* 1, as *De natura hominis*, ed. Moreno Morani (Leipzig: B.G. Teubner, 1987), p. 5.
88. Aquinas, *Summa theologiae* II-I q. 94 a. 2.
89. Aquinas, *Summa contra gentiles* III c. 3. See also *Summa theologiae* II-I q. 92 a. 1.
90. Ibid. q. 17 a. 1.
91. See Aristotle, *Topica* 1.1 100a–b.
92. *Summa theologiae* II-I q. 94 a. 2; q. 91 a. 3. See above p. 80.
93. Ibid.
94. E.g., Rhonheimer, *Natural Law*, pp. 11–12, 59, 63, 138.
95. Aquinas, *Commentarius in psalmos* ad loc.
96. *Summa theologiae* II-I q. 94 a. 2.
97. *Allen and Greenough's New Latin Grammar*, ed. J. B. Greenough et al. (Boston, 1931); *Medieval Latin: An Introduction and Bibliographical Guide*, ed. F.A.C. Mantello and A.G. Rigg (Washington, 1996).
98. Grisez, "First Principle," pp. 168, 174, 190–92, 181–82, 186, 182–83.
99. Note 87.

100. *Summa theologiae* II-I q. 17 a. 1. For "reason commands," see Rhonheimer, *Natural Law*, pp. 11–12, 59, 63, 138.
101. II-I q. 17 a. 1.
102. Ibid. q. 19 a. 3.
103. Ibid. q. 17 a. 1. Cf. I q. 19 a. 3.
104. Ibid. III q. 78 a. 2.
105. Ibid. II-I q. 90 a. 1. Aristotle, *Physica* 2.9 200a.
106. Note 72.
107. Ibid.
108. *Summa theologiae* II-I q. 17 a. 1.
109. E.g., Aquinas, *De veritate* q. 9 a. 7; *Summa theologiae* I q. 109 a. 3.
110. Ibid. II-I q. 92 a. 2.
111. Aquinas, *In Symbolum Apostolorum* a. 1; *Super Iohannem* 3.1.3; *Super Sententiarum* 4 d. 24 q. 1 a. 3 qc. 2 ad 3.
112. Charles E. Little, "The Authenticity and Form of Cato's Saying '*Carthago delenda est*,'" *Classical Journal* 29 (1934): 429–35. See also Silvia Thürlemann-Rapperswil, "'*Ceterum censeo Carthaginem esse delendum*,'" *Gymnasium* 81 (1974): 465–75. Pliny, *Historia naturalis* 15.20.74; Plutarch, *Cato maior* 27.2.
113. *Summa theologiae* I q. 79 a. 13.
114. II-I q. 17 a. 1. See above pp. 80–82.
115. *Summa theologiae* II-I q. 19 a. 5.
116. Ibid. q. 77 a. 2.
117. Ibid. a. 3; q. 90 a. 4; cf. Justinian, *Digestum* 44.7.3.
118. *Summa theologiae* II-I q. 94 a. 3.
119. Ibid. q. 106 a. 1.
120. François Garnier, *Le language de l'image au Moyen Âge: Signification et symbolique*, 2 vols. (Paris: Léopard d'or, 1982–89), 1:170.
121. Isidore of Seville, *Etymologiae* 11.1.66.
122. *Summa theologiae* II-I q. 95 a. 2.
123. Ibid. q. 91 a. 6.
124. Albert the Great, *De animalibus* 6.1.4, 12.2.1, ed. Hermann Stadler, 2 vols. (Munich: Aschendorff, 1916–20), 1:453, 838.
125. *Summa theologiae* II-I q. 91 a. 6.
126. Ibid., I q. 75; II-I q. 91 a. 1; I q. 81 a. 2; II-I q. 94 a. 2.
127. Leonard E. Boyle, *The Setting of the "Summa theologiae" of Saint Thomas* (Toronto: Pontifical Institute of Mediaeval Studies, 1982). See also Marian Michèle Mulchahey, *"First the Bow Is Bent in Study": Dominican Education before 1350* (Toronto: Pontifical Institute of Mediaeval Studies, 1998).
128. *Summa theologiae* II-I q. 94 a. 2.

129. See Torrell, *Aquinas*, pp. 4–7. Guillaume de Tocco, *Ystoria*, pp. 102–3. For Naples, see Darlene Pryds, "*Studia* as Royal Offices: Mediterranean Universities of Medieval Europe," in *Universities and Schooling in Medieval Society*, ed. William J. Courtenay and Jürgen Miethke with David B. Priest (Leiden: Brill, 2000), pp. 83–92. For Aquinas's education in theology at the University of Paris, see Torrell, *Aquinas*, pp. 36–74. See in general Spencer E. Young, *Scholarly Community at the Early University of Paris: Theologians, Education, and Society, 1215–1248* (Cambridge: Cambridge University Press, 2014).
130. Leonard Boyle, *Setting*. See also Mulchahey, *"First the Bow."*
131. *Summa theologiae* praef.
132. Torrell, *Aquinas*, 1:142–43, 146, 328–29; Torrell, *Aquinas's "Summa,"* pp. 10–15.
133. Torrell, *Aquinas*, 1:145–46, 160–61.
134. Leonard Boyle, *Setting*, pp. 26, 29.
135. Marjorie O'Rourke Boyle, "Chaff: Thomas Aquinas's Repudiation of His *Opera omnia*," *New Literary History* 28 (1997): 385. See now the critical edition of Guillame de Tocco, *Ystoria*.
136. Note 1.
137. See Emery, "Brief Chronology," p. 340.
138. Thomas Aquinas, *Super epistolam B. Pauli ad Romanos lectura* c. 2 lectio 3. It is dated to Naples 1272–73, and *Summa theologiae* II-I to Paris 1271. Emery, "Brief Catalogue," *Works*, pp. 340, 333.
139. Origen, *In Epistulam ad Romanos* 2.9.3, trans. Rufinus.
140. Marjorie O'Rourke Boyle, "Aquinas's Natural Heart," *Early Science and Medicine* 18 (2013): 266–90.

Jean Calvin, Heart in Hand

For his personal seal the Church reformer Jean Calvin (1509–1564) designed the heart in hand. As he wrote to Guillaume Farel, who by main force detained him in Geneva for the ministry, "My heart as if butchered I offer to the Lord in sacrifice."[1] Calvin's magisterial *Institutio christianae religionis* taught the biblical duty of believers to "present their bodies as a living holy sacrifice acceptable to God."[2] Calvin sacrificed his heart, for he was a cardiocentrist amid the medical and philosophical debates about whether the heart or the brain was the bodily seat of the soul. Although he was not a physician, he was a patient with chronic illnesses; and, although he was not a philosopher, he was a preacher who appropriated knowledge of nature and of medicine. Calvin thought the human body was a wonderful microcosmic creation deserving intense anatomical study. He urged his congregation to "weigh, with Galen's skill, its articulation, symmetry, beauty, and use."[3] Calvin's psychology was classical in origin as consistent with the prevailing Aristotelian doctrine that the heart was the seat of the soul. The heart operated both body and soul by its natural, meaning intrinsic, movements of physical contraction and expansion.[4] Calvin's psychology was also moralistic in its exegesis of the biblical Old Testament, where *lēb/lēbab* was its primary anthropological word and concept.[5] However, that Hebrew "heart" had no historical reference to the physical heart as an internal organ. Biblical heart was the agent that moved to or from observance of the divine law, which law defined Judaism as a

M. O'R. Boyle, *Cultural Anatomies of the Heart in Aristotle, Augustine, Aquinas, Calvin, and Harvey*,
https://doi.org/10.1007/978-3-319-93653-6_4

religion. Its physical metaphors for response to that law were the legs and feet, arms and hands, as the observable external limbs.[6] Calvin acknowledged that the biblical divine law resided in and possessed the sound and devoted heart that meditated on it.[7] He preached the necessity of the heart keeping that law wholly, without reservation for idolatry, which would corrupt and debase its service.[8]

Calvin's learning in biblical Hebrew produced relatively faithful translations.[9] Yet, his exegesis was in the Christian tradition that interpreted its "heart" anachronistically by philosophy and medicine on the operations of the physical heart in body and soul. His medical references, however, served theology not science. Calvin did not speculate medically on the modern proof text for a biblical cardiac knowledge, the story of the fool Nabal whose heart became stony and died. He preached on it morally in four exhaustive sermons that upbraided the injustices of Nabal's gross wealth then cautioned against the "cheerful heart" of his drunkenness.[10] Calvin's only notable resort to medical texts was second-hand, from his personal physician Benoît Textor, who had a book in progress on cancer and advised Calvin from detailed Roman sources that in 2 Timothy 2:17 the correct reading was gangrene, not cancer. Calvin's exegesis of that biblical disease was moral, the "horrible extinction of the gospel in the papacy" by the pastoral ignorance or inertia that corrupted doctrinal purity.[11] His ministry was to expound and preach Scripture toward the spiritual restoration of the pure heart, the seat of the soul. Calvin interpreted biblical "heart" generally as "the whole soul."[12] He eschewed, as obscure, philosophical subtleties. He preferred where apposite to divide soul simply into appetite, as will and concupiscence, and intellect, as theoretical and practical.[13] He thus interpreted "heart" as specifying in certain biblical verses the soul as the seat of the affects, in others its intellective part.[14] He relied on the Psalms as a full dissection, "an *anathomia* of all parts of the soul, for there is no affect in oneself that anyone can find that does not reflect in this mirror."[15]

The Cloven Heart

Calvin was essentially indifferent to precise philosophical divisions of the soul because he believed its faculties were incapacitated by the Fall. That was the original sin, a traditional Christian interpretation of Adam and Eve's disobedience of God's command by eating the forbidden fruit in the garden of Eden (Genesis 3).[16] Calvin defined original sin as "an inherited depravity and corruption of our nature diffused in all parts of the soul."[17] By it "soundness of mind and rectitude of heart were obliterated."[18] Original sin was no mere privation of Adam's primordial righteousness, as Calvin claimed most theologians maintained, but a relentless active depravity that schemed evil. Such concupiscence polluted the entire human being, the soul with its divisions of intellect and will, and even the body so that "the whole human is from himself nothing but concupiscence."[19] Sin possessed all parts of the soul, as a lawless impiety occupying the citadel of the mind to blind it, and as pride penetrating within to pervert the heart.[20] All natural faculties of the soul were so "vitiated and corrupted that prominent in all actions are perpetual disorder and intemperance."[21] The perverted heart was not Calvin's novelty. As the prophet Jeremiah had written, "The heart is deceitful above all things, and desperately corrupt" (Jeremiah 17:9). However, its deceit and corruption was not traceable in the Hebrew Bible to original sin, a distinctly Christian doctrine. Calvin interpreted that prophetic oracle historically as a counsel to the Jews against reliance on the heart since God alone searched its secrets. Calvin regretted that many other exegetes misevaluated that verse on the perverse and false heart as a pretext for reliance, instead, on the intellect.[22]

For the intellect Calvin's commentaries and sermons explained the idiom of speaking in the heart as "to ponder or conceive an opinion."[23] Communing with one's heart in leisure was like withdrawing to a high recessed bed to lie down to reason. "For solitude conduces people to connect with themselves, to examine themselves deeply, and to converse with themselves seriously about free decisions."[24] Calvin interpreted the frequent biblical sense of heart as "mind or understanding," or the soul distinguished as "reason, intellect, and will." But he thought there was never an understanding without an affect. Indeed, another meaning than "understanding" could be adduced because "the true knowledge of God is not imaginative, as they say, insofar as it is joined with serious affect." Thus, although Hebrew *heart* could be taken for mind, it also denoted "the will or the seat of the affects."[25] Calvin's theological commitment was to that faculty "for the heart bears the affections, the desires, the volitions: it is one thing to think something, and to desire it and to be devoted to it with a cordial affection."[26] He concentrated on the will because he believed that election for salvation belonged to it rather than to the intellect.[27]

Calvin's personal seal was the heart in the hand. His initial seal depicted on a shield flanked by his own Latin initials, "I C," a hand with its palm exposed. Between its thumb and index finger it poised a heart. His revised seal depicted on a cartouche the back of a hand with the heart in the same pose.[28] The emblem of the heart in hand also graced the border of Calvin's earliest engraved portraits.[29] The design was symbolic, as Calvin explained the union of those bodily parts. The biblical blessing on "clean hands and a pure heart," which inspired Jesus's beatitude, meant that "without a doubt the individual actions of our life need the heart's consent."[30] The hand was a commonplace metonym for power and symbolized action.[31] The handing of the heart to God was a motif of Renaissance art in personifications of Charity. In Giotto's invention Charity extends a bowl of fruit in one hand while she reaches above her head with her other arm. Open-handed with palms up and splayed fingers, she offers her heart to God. She holds it securely at the base, from which its severed aorta protrudes. She points the apex of her heart topsy-turvy up toward God, who accepts it fully in both his hands. Its reverse position of the heart on Calvin's personal seal of its apex below was introduced later, usually as an attribute of saints, notably Augustine and Aquinas.[32]

Those artistic hearts were shaped like pinecones, after their classical and medieval anatomical descriptions. They were broad at the "base," the modern wide top, and rounded at the "apex," the modern pointy bottom. The heart was for the first time depicted with a dented base in the mid-fourteenth century in Guido da Vigevano's *Anathomia designata per figuras*. Although that surgeon repeated the cardiac descriptions of the Arabic philosopher and physician Avicenna and of the medical authorities Henri de Mondeville and Mondino de' Luzzi, his illustrator deviated. No anatomical text had mentioned a cardiac dent. The medieval scholastic Albert the Great, whose paraphrases on Aristotle's animal books introduced his natural philosophy to Latin scholars, specified that the cardiac base was "undivided." But Aristotle had described the heart of large mammals, such as humans, as three-chambered. Between the two traditional ventricles he posited "an odd one in the middle." The Roman physician Galen corrected Aristotle's error of the third ventricle to a dilation or opening of the right ventricle at the base. Yet, despite Galen's greater authority in medicine, his observation faded when Avicenna favored Aristotle's description by inserting a third cavity between the two ventricles. For Avicenna, that cardiac cavity was a receptacle for bloody nourishment and for generation of the spirit from the airy refinement of the blood. Accuracy was further compromised by the classical terminology for the "base" and the "apex" of the heart, which authors and artists commonsensically reversed. Manuscripts with garbled descriptions, recopied by thoughtless scribes, compounded the confusion. And so the symbol of the heart deeply scalloped at the base, indented at tapered sides, and sharply pointed at the apex was invented in error.[33]

The heart on Calvin's seal had that erroneous shape. The design persisted, well after his physical heart ceased beating, to the modern icon ♥. Calvin's emblem does not hold the heart fully in hand, securely in its palm, as if like Charity he truly possessed his heart to offer to God. He holds his heart gingerly between his thumb and index finger, the joint signifier that distinguished humans from all other animals, for only humans have an oppositional thumb for grasping. What the human hand naturally grasped in Calvin's theology was sin, in perpetuity of Adam's original grasp of the forbidden fruit in the biblical garden of Eden.[34] Calvin's emblematic heart is poised precariously between grasping fingers at its anatomical apex but symbolic depth. The heart is divided: at its base the right and left ventricles are cloven. "A pious breast," wrote Calvin, "senses in itself a division, because part is imbued with sweetness on account of a recognition of the divine goodness, part is distressed with bitterness on account of a sense of its own calamity; part relies on the evangelical promise, part bristles with the testimony of its own iniquity; part exults in the apprehension of life, part is horrified at death." That variance derived from an imperfect faith in conflict with the flesh.[35] As Scripture warned, waffling with God was futile because he opposed "a double heart."[36] Calvin's cardiac motif abominated the double heart and valued the whole heart. Integrity versus hypocrisy was his thematic antithesis.[37] He exposed the division between inner heart and outer deceit,[38] as hypocrites knew their hearts were obscene but imagined that God would not spurn their deeds.[39] "The human heart has so many nooks of vanity, it abounds in so many lurking places of lies, and is so veiled with fraudulent hypocrisy, that it often deceives even itself."[40] Indeed, hypocrisy so ruled human nature that hearts became "a lot of mimes and monkeys."[41] Calvin observed that everyone regarded "cleanness of heart the mother of all virtues" but hardly anyone did not substitute "craftiness."[42]

Calvin stated that the true rule of pious living was "integrity of heart" for God hated nothing more than pretense.[43] Only a true and living faith purified a corrupt heart.[44] He acknowledged the Creator of all hearts from whose gaze no one could hide even in the recesses of the mind.[45] The heart was the conscience in the most intimate recess of the soul, where inward sanctity, rather than external appearance, pleased God.[46] A whole heart was intact and healthy. Explaining the biblical law to love and worship God "with a whole heart," he wrote that "a whole heart is taken for a sincere heart and is opposed to a double heart."[47] Again, "A whole heart is opposed to double or bisected one, and thus means the same as sound or minimally false."[48] The commandment "with the whole heart" meant "all affects of the heart."[49] Disposition to God meant pure affect, not a double heart,[50] a spiritual, not a sensory observance of the law.[51] As a pastor, Calvin was concerned to explain the liturgical significance of "with the whole heart" as "true worship in the interior affect of the heart," or "the sincere worship of God." Calvin contrasted true and sincere worship with "the external signs of the hands and feet, which are worthless toward perfection unless coming from a sincere heart."[52] For, God execrated nothing more than the counterfeit offering of external appearances for an innocent heart.[53] Liturgy should be celebrated "with sincere affect of the heart." The inner praise of God without witnesses was more excellent than public shouting with a feigned spirit and full-throated sounding. Hypocrites falsely exercised their tongues in the divine praises from "a cold and double heart."[54] Calvin praised the upright heart versus the hypocritical tongue, the ready heart versus lip service.[55] The double heart was double tongued since the Hebrew word for flattery meant "division."[56] He esteemed the "rare virtue of continence of the heart and tongue."[57] "Perfection of heart" meant an unsimulated worship.[58] "Certainly in the cult of God interior sincerity of heart occupies the first places."[59] Prayer and hypocrisy were so contradictory that not a word or a note profited anything unless coming from "a lofty affect of the heart."[60] All duplicity should be purged by the intrinsic motion of the Holy Spirit "so that an altar is erected to God in the heart itself."[61]

The Malicious Heart

Calvin denied that the construction of that cardiac altar was within human competence or ability. He believed that there was no human act toward salvation unless God intervened to create hearts anew.[62] For, ever since Adam's fall in original sin, all human cupidities were "evil."[63] Calvin characterized the human heart as evil in thought and deed. Its evil ruled pervasively as an infixed "malice," a depravity incurable by any ordinary remedy. This malice was not, he emphasized, an inclination to evil as the remnant of a mind originally created sound. It was a deeply imbued native character—a genius for malice from a damnable mind. The heart was not merely "vicious"; it had "not a drop of good" mingled in it.[64] He cited the biblical condemnation of the heart as a lucid mirror of human nature "fraudulent above all things and perverse." Although not every flagitious deed surfaced in every person, the hydra, the classical many-headed monster, swarming with vices lurked in all breasts. Just as the body fomented the matter and cause of disease and, even if not feverish with pain, it could not be called healthy, so the soul when it swarmed with its own diseases as vices could not be diagnosed as healthy.[65] Calvin inveighed on the corrupt cupidities of the heart steeped in sin and stinking.[66]

That stench moralized the putrefaction in the human body of worms, the prototypical unclean animals, which not only infested the stomach and intestines but also menaced the heart. In medical diagnosis, as worms twisted their way up to the heart, their fumes caused fevers, weakened and failed pulses, and mental illnesses. The pain and palpitation of the wormy heart upset the body with dizziness and with epileptic seizure and convulsion. Calvin thought the worms that nibbled metaphorically at the hearts of the faithful were "formed out of the earth." They were fleshy agents that caused "an awful and horrible fear and restlessness."[67] From such pollution Calvin believed that "everything that the human heart forges will always be perverse and malicious."[68] The heart only pretended that it was not corrupt. "A man forges in himself and in his workshop (French *boutique*) that he is not perverse and corrupt before God."[69] That workshop (Latin *officina*) was the heart as the fabricator of fictions.[70] Calvin's cardiac workshop or forge of evils moralized the artisanal metaphors of medical theory. The kindled hearth and the lit oven focused Aristotle's natural philosophy of the heart as the vital source of heat in blooded animals.[71] *De corde* (On the heart), which circulated later in the "Hippocratic" collection, imagined the heart as a burning hearth, with "ears" (the auricles) that functioned for cooling it much as smith's bellows was used in metal working.[72] The physical heart as the central workshop (*officina*) of the body would figure decisively in the seventeenth century for William Harvey's discovery of its circulation of the blood.[73] Calvin's heart as a furious workshop was remote from Augustine's heart as a leisurely bedroom (*cubiculum*), where he withdrew to listen with the cardiac ears to God's word and sang back to him love songs.[74]

Scripture afforded Calvin ample verses to comment and preach on the evil heart.[75] The first biblical mention of "heart" was of God's own heart grieving over his creation of humans because of the numerous and persistent evil thoughts and deeds of their hearts. His divine heart was sad because he no longer recognized humans as created in his own image and likeness.[76] Calvin believed that since Adam's original sin the biblical command to "love God with the whole heart" did not imply a human ability to comply but the very opposite "because our nature is entirely contrary to God." Calvin judged mistaken the papist inference from the divine commandments that humans could dispose themselves to obey them. He thought the commandments simply showed that humans were held to their observance. Calvin taught that everything humans were able to attempt was pointless and impure. Because they were incapable of any good movement toward God and in need of total renewal, it was God's intervention to "engrave his law in our hearts and entrails so that we hold to what he approves, which will be a conformity and agreement in all our desires and affects with this justice contained in the law."[77] At the ordinary Sunday service in Geneva the congregation sang the first four commandments, then the minister prayed that they would be "written on our hearts."[78] The engraved hearts on Calvin's personal seals and portraits[79] did not have any engraved letters. They were blank hearts, for divine engraving. As the motto on the portraits stated, *prompte et sincere*, ideally "ready at hand and whole" but not really.

Calvin characterized human hearts as naturally not only divided but also hard. Self-examination revealed "hearts of stone, in which not only hardness, not only malice reign but also that there is simultaneously an obstinacy that cannot of itself bend at all to obey God."[80] Although the popular design of the cloven heart was anatomically erroneous, there was medical consensus for the heart as muscularly hard. Comparative dissections of animal hearts reported their dense "hard flesh."[81] With that medical comparison Calvin's designation of the heart as hard acquired a significance beyond the biblical sense of a moral obduracy to the divine law. In the medical context Calvin meant that a hard heart exhibited its natural quality. In his theology hardness of heart signified "any contempt whatsoever for the word of God"—or even neglect of it—whether cold and contemptuous, slackish or squeamish, proud or furious.[82] "By nature we now have a certain stony heart, and this inborn hardness, which God alone is able to soften and correct, is in everyone from the womb." For that reason, humans spontaneously rejected the divine word. Everyone witnessed themselves as "authors and masters of their own inflexibility," with no one else to blame. "The cardiac faculty of forming obedience to God is hardly in our power. The heart is hardened until replaced from heaven."[83] The thoroughly depraved heart was impenetrable by the divine word until the Spirit softened its iron or stone. A contrary law was engraved there because perverse affects ruled the heart, forcing it to rebel. The preaching of Scripture was in vain until the Spirit inscribed his law in hearts. Calvin concluded against moral freedom of choice. "It is plain what *liberum arbitrium* is worth and what natural rectitude might be before God regenerates us." However humans might will or choose the good, they were swept away by a furious impulse toward opposing God and in no way capable of submitting to his justification. Thus the law was "fatal and moribund" if it remained written on tablets of stone. God had to "change and correct by his Spirit the native depravity of hearts."[84] Even with the gift of the Spirit to surmount all evil affects, believers groaned in base humility, aware of their weakness.[85] "What can a miserable manikin do when the softness of heart that is necessary to obedience is denied him? Yes indeed, what except shuffle his feet since his hardness can only be imputed to no one but himself?"[86]

Calvin addressed formally the topic "how God operates in human hearts."[87] He interpreted the frequent biblical verses that God himself "hardens the reprobate, turns, inclines, and forces their hearts." Passively, the Spirit was withdrawn from all humans at Adam's fall so that "our hearts harden into stones." Actively, the Spirit also hardened particular hearts to deliver them to ruin and destruction.[88] Calvin acknowledged the divine hardening of the heart as a very severe biblical saying, which many exegetes labored to mitigate to a mere permission. It was not necessary to write a dissertation, he thought, on how God hardened reprobates every time the biblical phrase occurred. It meant that God withdrew the grace of his Spirit from, and gave over to Satan, those whom he knew deserved it by their blindness of mind and obstinacy of heart. God was not the author of evil or sin, however. There should be no confusion about it because "hardness (*durities*) is the sin of a human, but hardening (*obduratio*) is the judgment of God" on sin.[89] Calvin's exegesis of the hardening of pharaoh's heart to resist the Exodus of the Israelites from Egypt thus interpreted a narrative motive as a theodicy.[90] He also interpreted the hardening of Israelite hearts by their disobedience to divine law to be their own as God's enemies.[91] "Human hearts are impelled by a secret instinct so that they will nothing, do nothing except by his nod." He intended that belief to contradict papist *liberum arbitrium*, which deviated from the pure gospel to profane philosophy, whence derived justification by works.[92] Calvin rejected philosophical causality for any change of heart. He explicated that none of Aristotle's four causes agreed with human works as constituting salvation. Rather, its efficient cause was the mercy of the heavenly Father and his gratuitous love; its material cause was Christ, whose obedience acquired justification; its formal and instrumental causes were faith.[93]

A corollary of the hard heart was the fat heart of the psalmist's condemnation and of Isaiah's prayer.[94] Calvin piously warned that the destiny of each human being was known to God alone, so that no one should dare to pry into the mystery of predestination to election or damnation.[95] However, he identified as a heretic Miguel Servet for denying the distinct hypothesis of the Son from the Father and for confusing the dual natures of Christ; and so approved Servet's execution by the Genevan Council.[96] As Calvin refuted his errors, instead of the physician Servet, "we have the physician Jesus Christ."[97] Calvin's doctrine was God fattening evildoers for the kill at the brink of the grave, just as a farmer fattened his cows or pigs for market before he slit their throats.[98] The "fat hearted" were "greasy." Even when their consciences gnawed within, the wicked had a certain grossness occupying their hearts so that they were stupefied and even furious in their obstinacy.[99] Calvin's association of fat pigs with the obstinacy of hard hearts borrowed Aristotle's example of fatty swine having a heart that was hard and dense, which quality rendered those animals temperamentally dull.[100] Calvin compared monks debauched in every vice—even to keeping convents as their brothels—to "pigs fattened in sties."[101]

The Contracted Heart

Calvin's qualities of hard and soft resorted not only to biblical diction about "heart." He also complemented metaphorically the traditional physiology of the cardiac movements of muscular contraction and expansion. Calvin moralized cardiac movement as naturally evil in its contraction into the self. Only the divine Spirit working through Scripture expanded the heart spiritually. It was Scripture, through its extrinsic sensory impressions on the ears or eyes of its hearers or readers, that moved the heart affectively. A change from natural constriction to spiritual relaxation happened by the motive power of the Spirit penetrating the heart. As Calvin explained, in adversity the human heart "contracts in straits" without relief, even in prayer, unless it should recline on God's own breast.[102] Divine law did not require external observance but internal obedience "so that the heart must somehow dilate itself." The heart could not achieve that power of expansion "by its own movement," however. It was God who would dilate hearts to make them fit for observance. With his correction of its hardness and inflexibility, the heart then offered itself freely as no longer fixed in "contractedness." Those whom God shaped to that "width" would lack no powers but have the faculty to act with a right affect.[103] Calvin commented on the dilated heart as "breadth of heart," signifying a "quickness." He acknowledged traditional psychology that "a narrow and contracted heart signifies either mourning or weariness or displeasure, to which is opposed the expanded heart for the opposite affects." It was everyday experience, Calvin acknowledged, that among friends "our heart spreads itself, all senses are open, nothing is hidden there, nothing shut: no indeed, the entire spirit itself leaps up and transports itself to be exposed openly."[104] The heart that expanded to act truly with neighbors was the whole undivided heart.[105]

However, for Calvin, self-examination revealed "hearts of stone, in which not only hardness not only malice reign there, but there is simultaneously an obstinacy that cannot of itself bend anything to obey God." He must intervene to give "hearts of flesh that will be soft and pliable so that we may serve him."[106] The divine method was double: inwardly through the Spirit, outwardly through the word, meaning the Bible. As Calvin explained, "By the Spirit, illuminating minds, forming hearts in the love and labor of what is right, he makes them a new creature. By the Word, he excites them for desiring, seeking, pursuing the same renewal."[107] Against the hard heart, Calvin emphasized the necessity of the Spirit's grace, which "softens, bends, and directs our hearts within to obedience to God."[108] The affects of the heart needed ordering to God with a sincere constancy. Papist arguments for human free choice either way Calvin dismissed as silly. "For it is the proper work of God, by the intrinsic movement of the Spirit, to convert to himself human hearts."[109] The divine determination was remote from the papist doctrine of an interim movement of human liberty to accept or reject divine law. God gave his children a new heart and promised that his Spirit would make them observe his commandments in perseverance to the end. For, grace was not proffered for a free human choice of acceptance or rejection. Rather, "Grace is that which forms in the heart both the choice and the will, so that every good work that follows is its effect."[110] God accepted the rectitude of the hearts of repentant sinners, he sanctified their imperfections, and he justified them by his free indulgence, which goods they could not acquire by their own merit.[111] The Spirit opened "the entrance of the heart" for his word and the sacraments, whose signs would otherwise only strike the senses.[112] "By his power alone hearts are penetrated, affects are deeply moved, and an entry lies open."[113] Just as the sensory organs of the body served their purposes, "so is the work of the Spirit in our hearts for conceiving, sustaining, fostering, and stabilizing faith."[114] In the sacrament of baptism the heart was wholly filled by faith, by which Christ indwelled.[115] Calvin advised that Christ's seat in the heart should not be "intuited from a distance in faith, but received by our soul in an embrace, so that he may dwell in us."[116] The purpose of his indwelling in hearts by faith was their reformation by the Spirit.[117] Baptism cleansed and regenerated to "spiritually create new humans."[118] The Spirit was a pledge of love that certified their faith.[119] The elect the Spirit regenerated were so efficaciously governed by him that their new hearts followed him with inflexible affection.[120]

The Spirit acted through Scripture by freely softening, even melting, the heart.[121] It was the proper office of the divine word, the Bible, which ministers preached, "to heal and to soften."[122] The Spirit must create a new heart by grace, which "forms in the heart both the choice and the will so that every good work that follows is its effect."[123] In the conversion of hard human hearts God radically extracted the stony heart and implanted a heart of flesh.[124] As Calvin taught, "The proper work of God is to circumcise hearts, and to give them for stony ones fleshy ones, to inscribe his law on their innards, so that by innovating souls his teaching may be efficacious."[125] Circumcision of the heart was performed by God's own hand so that those he renewed might love him.[126] It meant "a purgation from all depraved cupidities" secretly within lest people glory in their deeds.[127] Saints with the wisdom of Solomon prayed for that renewal by an inclination of their heart to God, who had shown them its obstinacy in sinful rebellion. The psalmist also confessed the "impurity of every part of the heart" and a "spirit twisted awry in depravity." He acknowledged that its recreation would be solely God's gift, thus prayed for a "clean heart" and a "right spirit in his innards."[128] Calvin concluded that "the entire human heart, where it is considered in its nature, is twisted and perverse." But the psalmist proclaimed "the magnificent and unique work of God in the renovation of humans, and so he makes new creatures whole."[129] Calvin argued that, since God ascribed to himself alone the renovation of the heart, it was a "sacrilege" for anyone to arrogate any part of it whatsoever to himself.[130] From his own youthful experience of being softened to biblical docility from papist hardness,[131] Calvin enjoined, "Let us truly learn that it is God's choice to bend human hearts in both directions so that he may cast down the bold with terror or raise up the timid."[132]

Calvin taught that Christian doctrine was "not of the tongue but of life; nor is it simply apprehended by the intellect and memory like the other disciplines but is only received then where it possesses the entire soul, and finds its seat and receptacle in the intimate affection of the heart." Christian doctrine "must discharge itself into the breast, and transfer into mores, and transform us into itself."[133] Calvin demanded "the piety that infixes certitude in hearts." He dismissed as "preposterous" all expert and elegant theological disputations toward toppling a solid faith in Scripture.[134] He summoned those censors who banished a heartfelt reverence for Scripture to take up the gospel and be cauterized in their consciences.[135] Against rational argument he pitted spiritual testimony. The same Spirit who spoke through the prophets had to "penetrate our hearts" to persuade them of the divine commandments.[136] For it was a fact that "those whom the Spirit has inwardly taught firmly acquiesce to Scripture, and this is *autopiston*, not to be subjected to demonstration and reasonings."[137] That self-authenticated evidence of the Spirit's confirmation of Scripture in human hearts countered the scholastic Aristotelian demonstrative topics, the first principles of science.[138]

Calvin denounced scholastics as brainy rather than hearty, although Aristotelian scholastics were not cephalocentrists but cardiocentrists.[139] His characterization of scholasticism as brainy was not physiological, however—the organic brain versus the organic heart—but psychological. In traditional medicine and philosophy a strict distinction between the physical and the psychological did not hold because the seat of the soul was located in, and sometimes divided between, the brain or the heart. (The notable scholastic dissent was Aquinas's theory of the soul as the form of the body, with the heart as the mover of the soul but not its residence.[140]) Calvin rejected the subtleties of the "pigpen" who applied by "preposterous arguments" Aristotle's "frigid doctrine" to refute the immortality of the soul by binding its faculties, as organic, to the body. "It is far fetched indeed that the potencies of the soul that serve the body are enclosed in its functions."[141] Beyond general disputation about the relationship of soul and body, Aristotelian scholastics identified the intellect with its reason as the prime faculties of the soul. Calvin emphasized the affectivity. His criticism of scholasticism as brainy was for its logical method of reasoning toward an intellectual understanding. Such understanding he rejected as notional, not powerful.[142] He judged scholasticism an ineffective method because it ignored the personal transformation that he believed was the necessity of faith and the office of ministry. However, Calvin repeated a traditional argument that the soul, although uncontainable in any particular place, was nevertheless in the body "as if to dwell in a domicile."[143] Although he declined to quibble about exactly where in the body faith was located, he decided on the heart rather than the brain. "Let us pay attention to the truth that the seat of faith is not in the brain but in the heart." He explained his choice not by philosophical reasoning or medical observation but by linguistic usage. "Since the word *heart* is generally taken for a serious and sincere affect, I say that it is a firm and efficacious trust, not some bare notion."[144]

The Secure Heart

In Calvin's judgment scholastics erred because "it is insufficient for the mind to be illuminated by the Spirit of God unless by his power the heart is stabilized and reinforced." He claimed that scholastics defined faith as a bare and simple assent to knowledge, a definition that omitted "the heart in trust and security." Calvin interjected that "the firm and stable constancy of heart is the principal part of faith."[145] He extolled "practical knowledge" of the divine power as "undoubtedly more certain and more solid than idle speculation." Thus, "The pious sensibility perceives God most intensely present where it feels itself vivified, illuminated, served, justified, and sanctified."[146] Calvin insisted on a knowledge of God that not "with inane far-fetched speculation flits about so much in the brain but one that in the future will be solid and fruitful, if it is observed by us with religious rite and it takes root in the heart." The legitimate seat of Christ indwelling the heart established that "it is insufficient if we roll him on the tongue or flit him about in the brain."[147] Unlike a scholastic Aristotelian movement from wonder at a sensory impression to the knowledge of its cause, Calvin believed investigation of God should "hold genius suspended in wonder so that simultaneously it may profoundly influence an efficaciousness in the sense."[148] As contrary to the knowledge of God that he believed to consist in worship, Calvin decried playful speculative propositions about the existence or nature of God as "frigid."[149] Christian faith was "a doctrine not of the tongue but of life; nor is it simply apprehended by the intellect and memory like the other disciplines but is only received then where it possesses the entire soul and finds its seat and receptacle in the intimate affection of the heart." Calvin rehearsed the biblical topic of the lips versus the heart to advocate for a faith that was not superficial in notion but efficacious in feeling. He decried frigid sophists "content to roll [the gospel] on the tip of their lips," whereas Christian eloquence had "to penetrate the intimate affections of the heart, to seat itself in the soul, and to affect the whole human being."[150] The ministerial office was "to kindle the fire of the Spirit in hearts and to melt and purge, even burn up, the affect of the flesh, and to truly excite a fervent love of God so that all humans might be seized to heaven by its flame."[151] For, unless biblical teaching was received "with sincere affect of the heart," it remained literal, "frigid scripture."[152] That frigid temperature copied the Aristotelian physiology of the brain. As physically distant from the heart, the bodily source of heat, the brain was reasoned to be the coldest organ. It functioned as a counterbalance to the heart to cool it lest it combust.[153]

Yet, although humans should praise God for his "noble laboratory" of their bodies, Calvin observed that with foul ingratitude they smothered within them the signs of divinity.[154] He cited Cicero's history that not even the barbarians of antiquity lacked a memory for the existence of God as the seed of religion in the heart. "Therefore, from the initial boundary of the world, there was no city, then no household able to lack religion. In this fact there is a certain tacit confession that a feeling of the numinous is inscribed in all hearts." Wanton idolatry only proved that the absolutely powerful impression of the numinous was irradicable. Humans were "imbued with a persuasion (*persuasio*) about God, from which, as if from a seed, emerged a propensity for religion."[155] The divinity implanted this religious seed in all hearts, although experience testified that hardly one in a hundred fostered it, while no one cultivated it to maturity. Like the psalmist's fool who "says in his heart 'there is no God,'" all humans extinguished their natural light and deliberately stupefied themselves.[156] Instead of serving God with sanctity and integrity, to curry his favor they devised silly trinkets and scrupulous observances that were worthless. Yet necessity extorted even from reprobates the confession that "an intimation of the numinous (*sensus divinitatis*) is naturally carved on human hearts."[157]

Calvin's sense of the numinous was not Aquinas's natural law, however. Calvin dismissed Aquinas's biblical proof text, the divine countenance signed upon human minds (Psalm 84:3 Vulg.) as a false apologetic for the utility of religious images. For Calvin, participation in God was not naturally possessed by an imprint of the eternal law on human reason. Participation was not attained by any personal works but only divinely conferred in Christian faith through the Spirit's activity. "Thus through him we arrive at participation in God, so that we may somehow feel his vivifying power toward us."[158] In distinction to Aquinas's natural law,

Calvin taught a persuasion of the heart, not a proof of the mind. Calvin's doctrine was the persuasion of the existence of gods or God, as created in human hearts, toward religious cult. Aquinas's doctrine was the attraction of the end of goodness, as created in human reason, toward moral behavior. Calvin's native piety was religious sentiment; Aquinas's natural law was rational conformity.[159] Calvin reversed Aquinas's premise of a natural goodness to a natural corruption that did not originate from created nature but from original sin. Calvin qualified his predication of "natural" as "extraordinary and accidental, not substantial." It designated human depravity not as originating from an individual habit but from the universal inheritance, Adam's original sin.[160]

Calvin acknowledged a sense of morality created in primordial human nature. The history of imperial Rome witnessed to the distinction between observance of and contempt for right and law. That distinction between honorable and base deeds God also engraved on the minds of individuals, and he often confirmed it by the disposition of his providence. Yet, as Calvin reasoned, however admirable for their virtuous reputation, not only did the Romans not merit a reward but even more so they deserved punishment for contaminating "the pure goods of God with the pollution of their hearts."[161] Despite a religious sentiment, Calvin insisted that an unclean cupidity manifested what the heart generated when left to its own devices, and how deep and indelible brands of ignominy were burned into the body.[162] Despite a created intimation of the numinous, since original sin there was an impassible gulf between human reputation and divine judgment. The good proceeding from the mind was evil and inefficacious. Only the good proceeding from the heart mattered, and that good was solely the Spirit's work. The actions of naturally good persons originated in evil affects; therefore, "from the very impurity of the heart, as from its origin, they are corrupt."[163] All works of all sinners were polluted by their impure hearts.[164]

Calvin's commentary on a proof text for a natural law, Romans 2:14–15, expounded Paul's argument distinctly. "He proves that ignorance is asserted in vain by the nations, since they declare by their deeds that they have some rule of justice. For no nation at any time so shrank back from humanity that it did not hold itself in check within some laws." That spontaneous and unsupervised inclination to lawmaking "manifested without a doubt that some concepts of justice and rectitude, which the Greeks call *prolēpsis*, are naturally inborn in human minds." Although the nations lacked the written Mosaic law, "nevertheless by no means did they utterly lack notions of the right and equitable." Otherwise, they could not have distinguished between outrage, which they punished, and virtue, which they commended and rewarded. By their deeds the ancients "testify that there is written on their hearts discernment and judgment by which they distinguish between the equitable and the iniquitous, the honorable and the base." However, Calvin argued, such knowledge toward religious ceremony and good behavior did not prove either free will or natural law, as the papists deduced. "For Paul did not teach the subjection of our faculty to the observance of the law, since he speaks not of a power for fulfilling the law but of a notion." In that verse, as in some other biblical texts, Calvin interpreted *heart* not as "the seat of the affects" but mere "understanding."[165] That latter interpretation, with reference to Deuteronomy 6:4–5 on loving God "with the whole heart," Calvin reiterated often.[166] And that interpretation was decisive because, for Calvin, the mind with its notions did not matter: the heart with its affects did. Faith was not notional but affective, not noetic but sensory. The bare apprehension of God in the heart grounded some knowledge of his existence but it was better understood as only an intimation (*sensus divinitas*).

In his commentary on the Pentateuch, Calvin elaborated on Paul's remark on the law of the nations in Romans 2:14–15 to expose its status as hypothetical. Calvin so expounded it by casting his verbs in the subjunctive, not indicative mood: would, would, would, and might. As Calvin wrote, Paul "commends the law because it would teach nothing but what nature itself would state most to be certain and equitable, and by which experience would demonstrate to us nothing more useful or more desirable, nevertheless at the same time he advises what might be any reason for its observance."[167] However, the hypothetical office of created nature in collaboration with divine law depended on the preservation of that created nature. Adam's fall had ruined its integrity for all posterity, who by that original sin inherited its depravity. Calvin's exegesis concluded that Paul "demonstrates clearly enough" that everyone was far removed from the firm observance of the command to love God with the whole heart. Indeed, human effort was "damaged and weak, unless the love of God occupies all our senses." What experience showed, as Calvin corroborated Paul, was how the mind was diverted to vanity and the affects to depravity. In sum, human motions were "evil."[168] Natural law was only a hypothesis about universal dysfunctionality that served to indict all humanity of Adam's inherited original sin.

Calvin reinforced his interpretation in his commentary on the prologue to John's gospel. He again considered the two principal parts of light still resident in corrupted nature, the created seed of religion and the conscientious discernment of good and evil. But, he asked, what of their fruits? Religion degenerated into a thousand portentous superstitions, while conscience so perverted every judgment that it confused vice with virtue. As he decided, "Natural reason never directs humans to Christ." However prudently they were instructed for regulating their lives, however talented they were for excellence in the arts and disciplines, after Adam's original sin their natural ability "totally vanishes without fruit." There was no advantage in the faint afterglow of human intelligence because Christ's glory was darkened and the human mind was blinded pitch black. The human operation of reason before its divine regeneration testified that human beings were created not only to breathe but also to understand. But reason "does not attain, or indeed approach God, so that their entire intelligence is nothing but sheer vanity." Calvin concluded that there was no human act toward salvation unless God intervened to create hearts anew.[169] That intervention would not restore the human heart to its original creation with an impression of a natural law but remake the human heart as a new creation with an engraving of the divine law.

The Sealed Heart

That engraving was the Spirit's seal on the heart that confirmed the biblical promise. However much the majesty of Scripture procured a mental reverence, "it does not affect us until has been sealed through the Spirit in our hearts." Again, the word of God should not spin at the top of the brain but root in the heart.[170] "For, if the true intelligence of the mind is his true illumination, much more evident does his power appear in such a confirmation of the heart; for the diffidence of the heart is greater than the blindness of the mind, and to instruct the spirit in security is more difficult than to imbue the mind with knowledge." Appealing to Paul's second letter to the Corinthians, Calvin taught that the Spirit "administers a seal for sealing those very promises on our hearts, the certainty of which he has previously impressed on our minds, and he takes the place of an earnest for confirming and constituting them."[171] The first act of the Spirit of adoption, which the elected received as a seal on their hearts, removed their hardness. It replaced the heart of stone with a heart of flesh.[172] Calvin's verb "to seal" (*obsigno*) classically designated sealing a letter,[173] a use for his personal seal engraved with the heart in hand. Calvin's theology applied but transcended Aristotle's influential comparison of the soul's reception of the sensible but immaterial form of an object to the impression on wax of the design on a signet ring.[174]

The personal seal of the Spirit was the gift of faith, a privilege, not a nature.[175] It was "the kind of persuasion that requires no reasons, the kind of knowledge that agrees with the best reason." As Calvin asserted, "I speak of only what every one of the faithful experiences." They knew experientially that "the true faith is solely the one that the Spirit of God seals on our hearts."[176] With that essential seal Calvin indoctrinated even children. The catechism for the Church at Geneva affirmed that the heart was too inclined to either a deficient or exaggerated self-confidence to allow God's movement rather than its own. As the child responded to the minister, "Truly the Holy Spirit by his illumination enables us to understand those things that otherwise far exceed our grasp and he forms us to sure persuasion by sealing the promises of salvation on our hearts."[177] That certitude about Scripture, which surpassed human argument, even ecclesiastical consensus, Calvin pronounced "wonderful" for the beautiful disposition of its parts that so solidly confirmed hearts. Surpassing the exemplary orators and philosophers—Demosthenes and Cicero, Plato and Aristotle—Scripture "attracts, delights, moves, and enraptures. It affects you, so that it will penetrate your heart and possess your marrow."[178] Calvin appealed to traditional psychology to indicate how an apprehension of the good would attract, then delight, then move, then seize. Beyond ordinary sensory impressions on the soul in the heart, he believed that Scripture, whether heard by the ears or read with the eyes, also affected the heart but spiritually so. That spiritual causality happened by the Spirit's motive power. As Calvin wrote, Scripture "affects (*afficiet*)."[179] The verb meant to exert an influence on body or soul, which caused the states called "affects (*affectus*)."[180]

The history of the classical affects and the medieval passions had deliberated their cause: whether they originated from a sensory impression in a movement of the heart by contraction or dilation, then affected the mind; whether a mental even locally cerebral impression reverberated in the body; or whether body and soul simultaneously acted or were acted upon. Natural philosophy in the Aristotelian and Stoic traditions designated the heart the seat of sensation and therefore of the affects or passions of the soul, which derived from external sensory impressions on the heart.[181] Calvin was a passionate cardiocentrist. For the Latin translation of Greek *pathos*, he consistently preferred *affectus*, a term of classical rhetoric and law, to *passio*, the ecclesiastical neologism standard in medieval scholastic psychology.[182] Calvin discussed affect by the traditional divisions of joy and gladness versus sorrow and worry, the rage of furor and the stupefaction of distrust.[183] He understood them also traditionally as a blow or strike to the mind that originated in a sensory impression.[184] He compared them to concussions, as when astonishment constituted stupor and horror "in a blow to the head."[185] Calvin acknowledged that it was humanly impossible "not to feel various movements." But, he counseled that, although humans might be "struck (*concuti*)," they should not be "devastated."[186] He knew that affects agitated people, whose hearts needed instead the stability of God's word. There was an "antithesis" between "the firm resolution by which the human heart adheres, where it is ruled by the Spirit, and the restlessness by which it is agitated, snatched to and fro as long as it fluctuates among its affects." The heart needed to assent to what was right "lest it boil over in depraved cupidities." For, "The human heart is turbulent, distraught, and as if shattered into various pieces until God restrains it, collected to himself in a firm and favorable grasp of obedience." So much, Calvin repeated, for the papist value of freedom of choice![187] "Solid rectitude of heart is the sheer gift of God."[188]

Calvin's affective psychology incorporated cardiac physiology: a palpitating versus a steady cardiac rhythm-based traditional morality concerning the affects or the passions. Calvin's warning against the heart boiling over in cupidities also acknowledged it as the source in Aristotelian natural philosophy of bodily, thus psychic, heat.[189] For Calvin, unbelievers were in "constant fluctuation," fearful lest the gods made sport of them. Believers too might be frightened at impending danger, unless they were stupefied, but they acknowledged God as their guardian. The stability of the upright was confidence in God with a trust that erased all anxieties.[190] Calvin thus counseled against fear.[191] Fear was his personal affect as characteristic of the phlegmatic temperament. In the prevailing Galenic medical theory, health and illness were determined by a balance or an imbalance of the bodily humors. Those humors were the sanguine, the phlegmatic, the bilious, and the atrabilious, as the bodily compounds of the elemental qualities of hot, dry, cold, and wet. The cause of Calvin's chronic diseases was diagnosed medically as an excess of phlegm, the wet and cold humor. Phlegm caused his kidney stones, arthritis, catarrh, pleurisy, bronchitis, and pulmonary tuberculosis. Personal experience thus dampened his notion of humanity, which he typified as phlegmatic, corrupted with fear and sloth.[192] The humor for an optimal heart was the sanguine humor, wet and hot, since the heart was the fount of the nourishing blood and the source of vital heat.[193] Calvin did not possess that naturally in himself, but he sought it spiritually from God. He compared the human medical practice of humoral diagnosis and therapy to the divine care of sick souls. As he argued, although God loved everyone equally, he treated individuals differently. "Why? because we have our different spirits and our constitutions (*complexions*), the same nourishment to all humans, the same medicine to sick persons would not make sense." God applied his diverse therapies according to his knowledge of what was expedient for each one.[194]

The Cured Heart

Among the affects, Calvin favored those opposite to his phlegmatic temperament: joy versus torpor, and the cheer and merriment that opposed mourning and constriction.[195] The matchless fruit of faith was a joy of heart and serenity of mind different from the shame or stupor of those unbelievers disturbed by inwardly felt motions. No one rejoiced acceptably, Calvin taught, unless he reclined on the one God and rested his health in that divine hand. There were countless disturbances everywhere, the only remedy for which was to focus on God, for faith would calm them. Not only did the faithful rejoice inwardly but even their flesh shared in that joy. And not only did the hidden affect of the heart foster joy spiritually but even the visible tongue told how much believers gloried in God as the guardian of health. God took custody not only of souls but also of bodies, which were not immune from troubles and injuries.[196] Theodore de Bèze, his first biographer and his successor at Geneva, praised Calvin's writings as all the more admirable because he had "a body so naturally weak, so diminished by sleeplessness and extreme abstemiousness, and also subject to so many illnesses, that anyone who should see him could not help think that he was hardly alive."[197] Calvin named as physicians those who from the pulse of the vein or artery indicated the state of bodily health.[198] He appealed to the Lord as a physician of the soul.[199] "He is a physician: let us expose to him our wounds.... He is the knower of hearts and aware of all thoughts: let us hasten before him to pour forth our hearts."[200] God was the wisest physician, the only sufficient healer.[201] As Calvin compared practices, good physicians looked beyond symptoms to causes and addressed the part or the root of the illness.[202] They were not invasive but careful to treat the particular disease so as not to mutilate the body but respect its integrity.[203] "If a man comes for a consultation and says 'I have a headache here and here,' the physician regards its source and applies the remedies he considered suitable. The Holy Spirit as a good physician for all our spiritual vices does the same."[204] But God was no experimenter. Although the procedure of human physicians whose prescribed remedies failed was to try the opposite,[205] divine prescriptions were perfect. Just as different persons suffered from different diseases so that was no one cure for all, "the heavenly physician treats some more mildly, others he purges with harsher remedies." Yet he cares for the health of everyone. "No one is passed by exempt and untouched because he knows everyone to be sickly."[206]

Classical psychology had treated the affects of the soul practically but distinctly. Aristotelians advocated their moderation as morally neutral, while Stoics required their rational evaluation and judgment. Stoic doctrine rejected the affects as agitations originating in assent to an involuntary and irrational blow to the mind. It counseled a certain insensibility to them.[207] Calvin rejected that Stoic therapy and even the assimilation of its *apatheia* to virtue in Christian asceticism. He criticized the "new Stoics among Christians who disallow groaning or weeping and also consider sadness and worry a vice." As he advised, "But this iron philosophy is nothing to us, since our teacher and Lord not so much by his word but even by the example of his own passion has condemned it."[208] Calvin also adduced the model of the psalmist, whose heart was "heaped in every nook and cranny with woes." Yet, "His heart was not iron or benumbed by Stoic hardness against pains and troubles." The psalmist did not succumb to his terrors but shielded himself by faith from their onslaughts.[209] The apostle Paul wept "with a softness of heart that was more heroic than that iron hardness of the Stoics."[210] Not for Calvin either that Stoic "loftiness of heart (*altitudo cordis*)" that surmounted passions with equanimity. He decried it as "pride and haughtiness" and "pride or ambition."[211] He counseled believers afflicted in heart to reject Stoic loftiness and, rather, prostrate themselves before God for him to raise them up. For, God was near to the faithful when they were wasting away with lethargy in their hearts.[212] Amid worry and sorrow all believers as humans felt turbulence, unless they were stupefied, yet they trusted in God to relieve their anxiety. Calvin taught that the peace of the faithful did not consist only in the spirit but extended to the body. "I respond that the faithful are quieted in the body because they trust confidently that they are totally in God's care, so that by his protection not only the soul will be safe but even the body will be guarded."[213] The Spirit so diffused its power in every part of the human heart that it not only mitigated sadness in adversity but also rendered tribulations lovable like a sweet condiment.[214] In adversity the Spirit enlivened hearts, newly invigorating the dead.[215] Calvin applied the biblical simile of withered hearts. Just as mowed grass no longer sucked moisture from the ground through its roots, so the heart as if eradicated or cut open was destitute of its natural nutrients.[216] But the Spirit strengthened the faithful to embrace with their whole hearts the safety he promised.[217] "By his aid he renders us invincible against all contests of Satan, as insidious as they are violent."[218] With that assurance the faithful acquired "trust of heart."[219]

Calvin's doctrine of the sanctification of the heart depended for its import on Aristotle's definition of nature as the possession of an intrinsic movement.[220] Calvin would have studied Aristotelian natural philosophy at the Collège de Montagu in Paris, a renowned center of dynamics and kinematics.[221] His theology rejected any natural human movement toward grace and salvation since the corruption of original sin. When Calvin wrote of "natural corruption," again, he qualified his modifier "natural" as an extraordinary and accidental predication, not a substantive one. For natural corruption did not designate Adam's originally created nature, which was good, but the effect of his original sin, the inheritance of which depraved his posterity. In that usage "natural" distinguished a universal hereditary corruption from an individual personal habit.[222] Any movement toward grace and salvation was necessarily extrinsic to humans. But to the Spirit the movement was personally intrinsic. Calvin declared, "It is the proper work of God, by the intrinsic movement of the Spirit (*intrinsico spiritus motu*), to reverse the direction of human hearts to himself."[223] The Spirit as the author of regeneration acted by "his own personal force (*proprio vigor*),"[224] as he had once moved at the original creation.[225] Calvin elaborated on his belief with Ezekiel's biblical prophecy of a new heart. "Regeneration is like another species of creation," he explained. "Truly, if one compares regeneration with the first creation, it far surpasses it."[226] The Church was the community of the single-hearted.[227] Yet, Calvin sighed that in "this bodily prison" no one was quick to be holy. He advised believers to make what little progress they could, striving daily without despair until "the infirmity of the flesh" was sloughed off and they were welcomed into full community with God in the next life.[228] As the supreme good of their hearts, the faithful expected union with God in their resurrection from the dead. Their desire for that satisfaction daily inflamed their hearts.[229]

Notes

1. John Calvin, *Epistolae*, in *Opera quae supersunt omnia*, ed. Eduard Reuss, Eduard Cunitz, and Johann Wilhelm Baum, 59 vols. in 26 (Brunswick: C. A. Schwetschke, 1863–1900), 11:100. Reference is usually to this edition, by the volume and column numbers. Reference, where available, is to the critical edition of Calvin, *Opera omnia* (Geneva: Droz, 1992–), cited as Geneva, by the volume and page numbers.
2. Calvin, *Institutio christianae religionis* 3.7.1, 2:505, citing Rom. 12:1.
3. Ibid. 1.5.3, col. 43; 1.5.2, col. 42. Marjorie O'Rourke Boyle, *Senses of Touch: Human Dignity and Conformity from Michelangelo to Calvin* (Leiden: E. J. Brill, 1998), pp. 172–75.
4. Aristotle, *De partibus animalium* 2.1 647a, 2.10 656a. Marjorie O'Rourke Boyle, "Aristotle's Cardiac Vessel," this volume, pp. 1–34.
5. Hans W. Wolff, *The Anthropology of the Old Testament*, trans. Margaret Kohl (Philadelphia, Pa.: Fortress, 1974), p. 40 counting 858 occurrences.
6. Marjorie O'Rourke Boyle, "'In the Heart of the Sea': Fathoming the Exodus," *Journal of Near Eastern Studies* 63 (2004): 17–27; "The Law of the Heart: The Death of a Fool (1 Samuel 25)," *Journal of Biblical Literature* 120 (2001): 401–27; "Broken Hearts: The Violation of Biblical Law," *Journal of the American Academy of Religion* 73 (2005): 731–57.
7. Calvin, *Commentarius in harmoniam evangelicam*, 45:611; *Commentarius in libros psalmorum*, 31:382; 32:219; *Commentarius in Mosis reliquos quatuor libros (Harmonia in pentateuchem)*, 24:227; Calvin, *Sermons sur le Deutéronome*, 26:443.
8. *Sermons sur Deutéronome*, 26:433.
9. Max Engammare, "*Joannes Calvinus trium linguarum peritus*?: La question de l'Hébreu," *Bibliothèque d'humanisme et renaissance* 58 (1996): 37–60.
10. Calvin, *Homiliae in primum librum Samuelis*, 30:527–73.
11. Calvin, *Commentarius in epistolam Pauli ad Timotheum II*, 52: 368–69. See Benoît Textor, *De cancri natura et curatione: ex probatissimis quibus[que] autoribus, tum Graecis, tum Latinis* (Leiden: J. Tornaesius, 1550). Calvin's commentary dated 1548 according to Theodore de Bèze and Nicolas Colladon, *Au lecteur chrestien*, 21:68–69. Calvin dedicated to Textor his commentary on the Second Epistle to the Thessalonians in memory of the doctor's medical care of Idelette de Bure Calvin, his wife. *Epistolae*, 13:598.
12. *In psalmos*, 31:688. For "soul," see *Institutio* 1.15.2, col. 135; for its faculties, 1.15.3–4, col. 138.

13. *Institutio* 1.15.6–7, cols. 140–42; cf. 2.2.3, cols. 186–87; *In psalmos*, 32:249.
14. Calvin, *In evangelium secundum Joannem commentarius*, ed. Helmut Feld, 2 vols. (Geneva), 11–1:106; *Commentarius in Pauli epistolas: ad Galatas, ad Ephesios, ad Philippenses, ad Colossenses*, ed. Helmut Feld (Geneva), *in Philippenses*, 16:375.
15. *In psalmos* 31:15.
16. For the tradition, see Dieter Groh, *Schöpfung in Wiederspruch: Deutung der Natur und des Menschen von der Genesis bis zur Reformation* (Frankfurt: Suhrkamp, 2003).
17. *Institutio* 2.1.18, cols. 182–83.
18. Ibid. 2.2.12, col. 195.
19. Ibid. 2.1.8, col. 183.
20. Ibid. 2.1.9, cols. 183–84.
21. Ibid. 3.3.12, col. 442. See also 1.15.8, cols. 142–43; 2.2.1, cols. 185–86; 2.2.12, cols. 195–96; 2.2.16, col. 207.
22. Calvin, *Praelectiones in librum prophetiarum Ieremiae*, ed. Nicole Gueunier with Max Engammare, 2 vols. (Geneva), 6–1:658–59.
23. *Harmonia in pentateuchem*, 24:234, and see cols. 279, 280; *In psalmos*, 31:135–36, 483; Calvin, *Praelectiones in XII prophetas* minores, 44:21, cf. Calvin, *Sermons sur l'harmonie évangélique*, 46:301–02.
24. *In psalmos*, 31:61, 713.
25. Re: Deut. 6:4–5//Matt. 22:37 see, Calvin, *In evangelium secundum Johannem commentarius*, ed. Helmut Feld (Geneva), 2 vols., 11–2:106; *Harmonia in pentateuchem*, 25:61; Calvin, *Sermons sur le livre de Job*, 34:46; Calvin, *In Jeremiam*, 6–1:904; Calvin, *Commentarius in harmoniam evangelicam*, 45:442, 611; Calvin, *Commentarius in Acta Apostolorum*, ed. Helmut Feld (Geneva), 2 vols., 12–2:76.
26. *Sermons sur Deutéronome*, 26:434.
27. *Institutio* 2.2.26, col. 207.
28. Émile Doumergue, *Jean Calvin, les hommes, et les choses de son temps*, 7 vols. (Lausanne: Georges Bridel, 1899–1927), 1:569.
29. Émile Doumergue, *Iconographie calvinienne* (Lausanne: Georges Bridel, 1909), pls. XIV, XV, XVa, and see pp. 55, 64.
30. Ps. 24:4. Calvin, *Homiliae in primum librum Samuelis*, 30:167. See also *In harmoniam evangelicam*, 45:163.
31. Boyle, *Senses of Touch*, pp. 206–27.
32. Pierre Vinker, *The Shape of the Heart*, rev. ed. (Amsterdam: Elsevier, 2000), pp. 34–43. See also Anne Derbes and Mark Sandona, "'*Ave charitate plena*': Charity in the Arena Chapel," *Speculum* 76 (2001): 599–637. See the frontispiece of this volume.
33. Vinker, *Shape of the Heart*, pp. 44–62, 80–88.
34. Boyle, *Senses of Touch*, pp. 22–23, 72–76, 109–12. Genesis 3:6.

35. *Institutio* 3.2.18, col. 413.
36. Ibid. 3.3.6, col. 438; 3.6.5, col. 505; Calvin, *In Isaiam* 36:648.
37. *Institutio* 2.8.5, col. 304; 3.20.7, col. 631; 3.20.31, col. 658; 3.24.8, col. 719; 4.12.19, col. 917; *Sermons sur Deutéronome*, 26:434–35; 28:570; *In harmoniam pentateuchem*, 28:570; *In psalmos*, 31:207, 282, 557, 560, 733, 796; 32:217, 249; Calvin, *Commentarius in librum Isaiae*, 37:313, 641; Calvin, *Praelectionum in Ieremiam prophetam*, 37:641; Calvin, *Praelectiones in Ezechielis viginti capita priora*, 40:308; *In harmoniam evangelicam*, 45:131, 227; *In evangelium Ioannis* (Geneva), 11–1:182–83; Calvin, *Commentarius in epistolam Pauli ad Thessalonicenses 2*, 52:204; Calvin, *Commentarius in Iohannis apostoli epistola*, 55:342; Calvin, *In Ephesianos*, in *Commentarii in Pauli epistolas* (Geneva), 16:264. *In psalmos*, 31:207; 32:167; 31:675. Calvin, *Commentarius in epistolam Pauli ad Colossenses*, in *Commentarii in Pauli epistolas*, p. 453.
38. *In psalmos*, 31:421.
39. *Institutio* 3.14.7, col. 568.
40. Ibid. 3.2.10, col. 406.
41. *Sermons sur Job*, 33:544.
42. *In harmoniam evangelicam*, 45:163.
43. *In Isaiam* 36:648.
44. *Institutio* 3.14.8, col. 570; 3.20.5, cols. 301–2.
45. *In psalmos* 33:331.
46. *In Thessalonicenses 1*, 52:159.
47. *In psalmos*, 32:372; and see 32:349.
48. *Harmonia in pentateuchem*, 24:224.
49. *In psalmos*, 32:280; *Institutio* 2.8.51, col. 303.
50. *Sermons sur Iob*, 33:545.
51. *Institutio* 2.8.6, col. 270; *In psalmos*, 32:229.
52. *Institutio* 2.8.6, cols. 270–71; *Homiliae in librum 1 Samuelis*, 30:167; *In psalmos*, 32:216, 263; Calvin, *Praelectiones in lamentationes Ieremiae*, 39:592.
53. *Institutio* 4.12.19, col. 917.
54. *In psalmos* 32:167, 281. *Institutio* 2.8.5, col. 304; *In harmoniam pentateuchem* 28:570; *In psalmos* 32:249; *In Isaiam*, 37:313:641; *In Ieremiam*, 40:308; *In harmoniam evangelicam*, 45:131; *In Ioannem*, 11–2:182–83; *In Thessalonicenses 2*, 52:204; Calvin, *Commentarius in Ioannis epistola*, 55:342.
55. Calvin, *In psalmos*, 31:282, 557, 560, 733, 796; 32:217; *Institutio* 3.2.10, col. 405, 3.24.8, col. 719; *In harmoniam evangelicam*, 45:227, 343–44; Calvin, *Commentarius in epistolam Pauli ad Romanos*, ed. T. H. L. Parker and D. C. Parker (Geneva), 13:221; *Commentarius in epistolam Pauli ad Ephesianos*, in *Commentairii in Pauli epistolas* (Geneva), 16:264.

56. *In psalmos* 31:127.
57. Ibid. col. 207.
58. *In psalmos*, 32:167.
59. *In psalmos* 31:675.
60. *Institutio* 3.20.7, col. 631; 3.20.31, col. 658; *Commentarius in epistolam Pauli ad Colossenses*, in *Commentarii in Pauli epistolas* (Geneva), 16:453.
61. *Institutio* 3.3.16, col. 446–47.
62. *In Johanem*, 11–1:19–20.
63. *Institutio* 3.3.12, col. 442.
64. Ibid. 3.14.1, col. 564. Genesis 6:5. Calvin, *In Genesin*, 23:117–18; cf. col. 139.
65. *Institutio* 2.3.2, cols. 210–11.
66. Ibid. 2.2.12, cols. 195–96; 2.2.26, col. 207; 2.5.19, col. 247; 3.2.41, col. 432; 3.4.9, col. 523; 2.1.1, cols. 185–86.
67. Boyle, *Senses of Touch*, pp. 184–87.
68. *Sermons sur Deutéronome*, 28:528.
69. *Sermons sur Deutéronome*, 26:284.
70. *In psalmos*, 32:393.
71. See "Aristotle's Cardiac Vessel."
72. Hippocratic corpus, *De corde* 5, 8, in *Hippocrates*, Vol. 9, ed. Paul Potter (Cambridge, Mass.: Harvard University Press, 2010). For this work, see I. M. Lonie, "The Paradoxical Text 'On the Heart,'" *Medical History* 17 (1973): 1–15, 136–53; and relative to the Hippocratic collection, Elizabeth M. Craik, *The 'Hippocratic' Corpus: Content and Context* (Abingdon, Oxon.: Routledge, 2015), pp. 53–56.
73. Marjorie O'Rourke Boyle, "Harvey in the Sluice: From Hydraulic Engineering to Human Physiology," *History and Technology* 24 (2008): 1–2. See also "Aristole's Cardiac Vessel," p. 9.
74. Augustine, *Confessionum libri tredecim* 8.8.19,10.31.46, 12.16.23, ed. Lucas Verheijen (Turnout: Brepols, 1981); Augustine, *Sermones* 352.3.9, 179.7.7, in *Patrologiae cursus completus, series latina*, ed. Jean-Paul Migne (Paris, 1800–75), vols. 38–39. Marjorie O'Rourke Boyle, "Augustine's Heartbeat: From Time to Eternity," *Viator: Medieval and Renaissance Studies* 38 (2007): 40–41.
75. *Institutio* 3.14.1, col. 564.
76. *In Genesin*, 23: cols. 117–18; cf. col. 139.
77. *Sermons sur Iob*, 33:546; *Sermons sur Deutéronome*, 28:285, 562. For engraving on hearts, see also *Harmonia in pentateuchem*, 25:54; *Sermons sur Deutéronome*, 29:87; *In psalmos*, 31:823; *Commentarii in Secundam Pauli Epistolam ad Corinthios*, ed. Helmut Feld (Geneva), pp. 50–51.

78. T. H. L. Parker, *John Calvin: A Biography* (London: Dent, 1975), p. 86, citing *La Manyere de faire prières.*
79. See notes 28 and 29.
80. *Sermons sur Iob*, 33:546.
81. C. R. S. Harris, *The Heart and the Vascular System in Ancient Greek Medicine: From Alcmaeon to Galen* (Oxford: Clarendon, 1973), pp. viii, 75, 76, 173–175, 271, 432–455.
82. *In psalmos*, 32:32–33.
83. Calvin, *Commentarius in epistolam ad Hebraeos*, ed. T. H. L. Parker (Geneva), 19:53, 30. See also *Institutio* 2.3.8, cols. 217–18; 2.3.8, col. 218; 2.3.9, cols. 218–19; 2.5.4, cols. 232–33; 2.5.12, cols. 240–41; 3.24.14, cols. 724–25; 3.24.16, cols. 725–27; 4.1.6, cols. 751–52; *In Acta Apostolorum*, 12–1:219; *In Ieremiam*, 6–1:162–63; cf. *In Romanos*, pp. 54, 86–87; *Harmonia in pentateuchem*, 24:253; *In psalmos*, 32:230. For God's hardening of the heart and its hardness, see *Institutio* 2.4.3, cols. 225–26; *Harmonia in pentateuchem*, 24:86–87, 222; *In psalmos* 31:764; 32:109. *Sermons sur Deutéronome*, 26:375.
84. *In Hebraeos*, p. 30.
85. *Sermons sur Deutéronome*, 26:375.
86. *Institutio* 2.5.5, col. 233.
87. Ibid. 2.4 title, col. 224.
88. Ibid. 2.4.3, col. 266.
89. *Harmonia in pentateuchem*, 24:86–87.
90. *Institutio* 1.18.2, col. 155; 2.5.6, cols. 227–28; 3.24.13, col. 723.
91. Ibid. 2.5.11, col. 239; *In psalmos*, 31:764.
92. *In psalmos*, 32:109; and see *Institutio* 2.5.11, col. 239.
93. *Institutio* 3.14.17, cols. 575–76.
94. Ps. 17:10, 119:70; Is. 6:10.
95. *Institutio* 3.21.2, cols. 681–82.
96. For Servet, see Bruce Gordon, *Calvin* (New Haven, Conn.: Yale University Press, 2009), pp. 217–33.
97. Calvin, *Refutatio errorum*, 8:535.
98. *Sermons sur Iob*, 34:72; 35:197.
99. *In psalmos*, 32:245. For piggish objectors to divine predestination, see also *Institutio* 3.23.12, col. 708; cf. Aquinas at 22.9, col. 695.
100. Aristotle, *De partibus animalium* 3.4 667a.
101. *Institutio* 4.13.15, col. 936. See also "pigpen" Epicureans, 1.5.4–1.5.5, cols. 43–44.
102. *In psalmos*, 31:588–89, 61.
103. *In psalmos*, 32:227–28.
104. *In Corinthios 2* (Geneva), pp. 153, 113.

105. *In psalmos*, 31:127.
106. *Sermons sur Iob*, 33:546.
107. *Institutio* 2.4.5, cols. 233–34.
108. Ibid. 3.20.46, col. 673.
109. *In psalmos*, 31:725.
110. *Institutio* 2.3.13, col. 223.
111. *In psalmos*, 31:323.
112. *Institutio* 4.14.8, col. 947; 1.9.3, col. 71; 2.5.6, col. 235.
113. Ibid. 4.14.9, col. 947; 4.14.17, col. 954.
114. Ibid. 4.14.9, cols. 947–48; 4.14.19, col. 956.
115. Ibid. 4.14.8, cols. 946–47; 4.17.5, col. 1006.
116. *In Ephesianos* (Geneva), pp. 214, 215.
117. Calvin, *Sermons sur l'Epitre aux Ephesiens*, 51:489.
118. *Institutio* 4.14.4, col. 943; 4.14.24, col. 959; *Harmonia in pentateuchem*, 25:54.
119. *In Galatas* (Geneva), 16:94–95.
120. *Institutio* 2.3.10, col. 220–21; 3.20.43, cols. 667–68; *In Iohannis epistola*, 55:336.
121. *Institutio* 3.24.14, col. 724, 3.24.16, col. 726; *In harmoniam pentateuchem* 24:253; *In psalmos* 32:33.
122. *In Isaiam*, 36:139.
123. *Institutio* 2.3.13, col. 223; *In psalmos*, 31:323.
124. *Institutio* 2.3.8, cols. 217–18.
125. Ibid. 2.5.4, col. 233.
126. Ibid. 2.5.12, cols. 240–41.
127. *Harmonia in pentateuchem*, 24:222; *In Acta Apostolorum* (Geneva), 12–1:219; *In Ieremiam*, 37:587; cf. *In Romanos* (Geneva), pp. 54, 86–87.
128. *Institutio* 2.3.9, col. 218.
129. *In psalmos*, 31:518. Calvin typically cites that text to refute papist *liberum arbitrium*, or a moral freedom of choice toward grace.
130. *Institutio* 4.1.6, col. 752.
131. *In psalmos*, 31:21.
132. *Harmonia in pentateuchem*, 24:253; *Sermons sur Iob*, 33:546; *In psalmos*, 32:230.
133. *Institutio* 3.3.16, col. 446.
134. Ibid.1.7.4, col. 59.
135. Ibid. 1.8.11, col. 67.
136. Ibid. 1.7.4 col. 59.
137. Ibid. 1.7.5, col. 60.
138. See Thomas Aquinas, *Summa theologiae* II-I, q. 94 ad 2.
139. E.g., Marjorie O'Rourke Boyle, "The Wonder of the Heart: Albert the Great on the Origin of Philosophy," *Viator: Journal of Medieval and*

Renaissance Studies 45 (2014): 266–90; idem, "Aquinas's Natural Heart," *Early Science and Medicine* 18 (2013): 266–90.

140. Boyle, "Aquinas's Natural Heart."
141. *Institutio* 1.5.5, col. 44.
142. *In Romanos* (Geneva), pp. 46–47.
143. *Institutio* 1.15.6, col. 140. For "domicile," see, e.g., Alfred of Shareshill, *De motu cordis*, in *Des Alfred von Sareshel (Alfredus Anglicus) Schrift "De motu cordis,"* ed. Clemens Baeumker (Münster: Aschendorff, 1923), pp. 9–12.
144. *In Romanos* (Geneva), p. 221.
145. *Institutio* 3.2.33, cols. 425–46.
146. Ibid. 1.13.13, col. 101.
147. *In Ephesios* (Geneva), p. 214.
148. *Institutio* 1.5.9, cols. 47–48; 3.2.34, col. 637. For scholastic Aristotelian wonder, see Boyle, "Wonder of the Heart."
149. *Institutio* 1.2.2, cols. 34–35; 1.12.1, col. 86.
150. Ibid. 3.6.4, col. 504.
151. *In harmoniam evangelicam*, 45:809–10.
152. *In Romanos* (Geneva), p. 55.
153. Aristotle, *De partibus animalium* 2.7 652a–653a; *De generatione animalium* 5 743b; *Historia animalium* 1.13.2 945a.
154. *Institutio* 1.12.1, cols. 86–87; 1.5.9, col. 47; 1.10.3, cols. 73–74; 1.5.9, col. 48; 3.2.34, cols. 426–27; 1.5.4, cols. 43–44.
155. Ibid. 1.3.1–2, cols. 36–37.
156. Ibid. 1.4.2, col. 39.
157. Ibid. 1.3.1–3, cols. 36–38, 1.10.3, col. 74; *In Samuelem 1*, 30:36; *Institutio* 1.3.2, cols. 36–37; 1.4.1, cols. 38–39; 1.4.2, col. 39; 1.4.4, cols. 40–41. For another interpretation of *sensus divinitatis*, see Paul Helm, *John Calvin's Ideas* (Oxford: Oxford University Press, 2004), pp. 218–45. See also David C. Steinmetz, "Calvin and the Natural Knowledge of God," in his *Calvin in Context* (New York: Oxford University Press, 1995), pp. 23–39; Susan E. Schreiner, *The Theater of His Glory: Nature and the Natural Order in the Thought of John Calvin* (Durham, N.C.: Labyrinth, 1991), pp. 73–95.
158. *Institutio* 2.2.1, col. 186; 1.13.14, col. 102.
159. See Marjorie O'Rourke Boyle, "Aquinas's Law of the Heart: Natural Reason," this volume, pp. 67–100.
160. *Institutio* 2.1.11, col. 184.
161. Ibid. 3.14.3, col. 565.
162. *In Romanos* (Geneva), p. 35.
163. *Institutio* 3.14.3, col. 565.
164. Ibid. 3.14.7, cols. 568–69.

165. Calvin, *In Romanos* (Geneva), 13:46–47.
166. *Harmonia in pentateuchem*, 25:61; *Sermons sur Job*, 34:46; *In Ieremiam*, 38:462; *In harmoniam evangelicam*, 45:442, 611; *In Johannem* (Geneva), 11–2:106; *In Acta Apostolorum* (Geneva), 12–2:76.
167. *Harmonia in Pentateuchem*, 24:723. Calvin's clausal subjunctives are all mistranslated as indicatives in Calvin, *Commentaries on the Four Last Books of Moses Arranged in the Form of a Harmony*, trans. Charles W. Bingham, 4 vols. (Grand Rapids, Mich.: Wm. B. Eerdmans, 1950), 3:193.
168. *Harmonia in Pentateuchem*, 24:723.
169. *In Johannem* (Geneva), 11–1:19–20.
170. *Institutio* 1.7.5, col. 60.
171. Ibid. 3.2.36, col. 428; Calvin, *Le catechisme de l'Eglise de Genève/ Catechismus ecclesiae genevensis*, 6:45–46.
172. *Institutio* 3.24.1, col. 712.
173. "obsigno," *A Latin Dictionary*, ed. Charleton T. Lewis and Charles Short (Oxford: Clarendon, 1969), s.v.
174. See Aristotle, *De anima* 2.12 424b.
175. *Institutio* 3.2.42, col. 432.
176. Ibid. 1.7.5, col. 60.
177. *Catechisme/Catechismus*, 6:45–46.
178. *Institutio* 1.8.1, cols. 61–62.
179. Ibid.
180. "affectus," *Latin Dictionary*, s.v.
181. For an introduction, see Simo Knuuttila, *Emotions in Ancient and Medieval Philosophy* (Oxford: Clarendon, 2004). See also Boyle, "Wonder of the Heart" and "Aquinas's Natural Heart."
182. "affectus," "passio," *Latin Dictionary*, s.v. See Quintilian, *Institutio oratoria* 6.2.8. For Calvin on Aristotle's *pathos*, with evil *affectus* as *perturbatio*, see ibid. 2.2.2, col. 204.
183. *Harmonia in pentateuchem*, 24:58.
184. *In Genesin*, 23:557.
185. *In Isaiam*, 36:322.
186. *In Johannem* (Geneva), 11–2:135–36.
187. *In psalmos*, 31:795.
188. *In psalmos*, 32:249.
189. Aristotle, *De partibus animalium* 3.4 665b–666a; 3.7 670a; *De iuventute et senectute* 8 474a–b. See Boyle, "Aristotle's Cardiac Vessel."
190. *In psalmos*, 32:175.
191. *Harmonia in pentateuchem*, 27:597–99.

192. Boyle, *Senses of Touch*, pp. 250–55.
193. See "Aristotle's Cardiac Vessel."
194. *Sermons sur le Deutéronome*, 29:190. For Calvin's descriptive diagnosis of "the species of melancholy" in Saul's behavior, see *Homiliae in primum librum Samuelis*, 30:181.
195. *Harmonia in pentateuchem*, 24:58.
196. *In psalmos*, 31:156.
197. Théodore de Bèze, *Au lecteur chrestien I* (Vie de Calvin), 21:33.
198. Calvin, *Praelectionum in Danielem prophetam*, 40:561.
199. *Sermons sur le livre de Job*, 33:266; *In psalmos*, 31:75, 230; *In prophetas minores*, 42:337, 358, 504, 505; 43:142; *Commentarius in harmoniam evangelicam*, 45:156, 444; *In Romanos* (Geneva), p. 116.
200. *Institutio* 3.4.9, col. 464; Calvin, *Responsio contra Pighium*, 6:367; Calvin, *Articuli a facultate s. theologiae Parisiensi*, 7:11; *In psalmos*, 32:155. For the Eucharistic medicine, see Calvin, *De la cene*, 5:445; Calvin, *La forme des prieres et changz ecclsiastiques*, 6:199.
201. *Homiliae in primum librum Samuelis*, 29:395. *Commentariorum in quinque libros Mosis*, 24:164; *In Isaiam*, 37:258.
202. *In Isaiam*, 37:6; Calvin, *Praelectiones in duodecim prophetas minores*, 42:337; *Sermons sur la premiere epitre a Timothee*, 53:207. Cf. bad physicians who neglect the causes, *In psalmos* 31:253–54.
203. *In Corinthios 2* (Geneva), p. 215.
204. *Sermons sur la premiere epitre a Timothee*, 53:207.
205. Calvin, *In "De clementia"* 1.17; *Calvin's Commentary on Seneca's "De clementia,"* ed. Ford Lewis Battles and André Malan Hugo (Leiden: E. J. Brill for the Renaissance Society of American, 1969), p. 266.
206. *Institutio* 3.8.5, col. 518. *In Ieremiam* (Geneva), 6–1:251–53; *In prophetas minores*, 42:318; 44:496; *Commentarius in epistolam Pauli ad Timotheum I*, 52:305.
207. Richard J. Sorabji, *Emotion and Peace of Mind: From Stoic Agitation to Christian Temptation* (Oxford: Oxford University Press, 2000), pp. 194–210, 66, 67, 69. However, for eupathic responses, see Margaret R. Graver, *Stoicism and Emotion* (Chicago: University of Chicago Press, 2007). In general see also Knuuttila, *Emotions.*
208. *Institutio* 3.8.9, cols. 520–21.
209. *In psalmos*, 31:261, 387, 389–90, 398; 536. Ibid. cols. 272, 415. Ibid. cols. 580–81.
210. *In Corinthios 2* (Geneva), p. 37.
211. *In Isaiam*, 36:306–307.
212. *In psalmos*, 31:344; 32:339.
213. *In Acta Apostolorum* (Geneva), 12–1:71–72.

214. *In Romanos* (Geneva), p. 104.
215. *In psalmos*, 31:652.
216. *In psalmos*, 32:63.
217. *In psalmos*, 31:313; *Institutio* 3.2.7, col. 412.
218. *Institutio* 3.20.46, col. 673; *In Isaiam*, 36:139.
219. *Institutio* 3.24.7, col. 718.
220. Aristotle, *Physica* 2.1 192b; and see 1.18 19b.
221. See Marcel Godet, *La congrégation de Montaigu (1490–1580)* (Paris: H. Champion, 1912); Augustin Renaudet, *Préréforme et humanisme à Paris pendant les premières guerres d'Italie (1498–1517)*, 2nd rev. ed. (Paris: Librairie d'Argences, 1953), pp. 267–72; Hubert Elie, "Quelques maîtres de l'Université de Paris vers l'an 1500," *Archives d'historie doctrinale et littéraire du Moyen Age*, 18–19 (1950–51): 222–24; Olivier Millet, *Calvin et la dynamique de la parole: Étude de rhétorique réformée* (Geneva: Slatkine, 1992), pp. 30–34. See also Christopher B. Kaiser, "Calvin's Understanding of Aristotelian Natural Philosophy: Its Extent and Possible Origins," in *Calviniana: Ideas and Influence of Jean Calvin*, ed. Robert V. Schnucker (Kirksville, Mo.: Sixteenth Century Journal Publishers, 1988), pp. 77–92.
222. *Institutio* 2.1.10–11, cols. 184–85.
223. *In psalmos*, 31:725.
224. *Institutio* 1.13.14, col. 102. Against the identification of God as nature, see col. 45.
225. Genesis 1:2. *In Genesin*, 23:11, 16.
226. *In Ezechielem*, 40:456.
227. *Institutio* 4.1.3, cols. 747–48.
228. Ibid. 3.2.18, col. 413; 3.3.6, col. 438; 3.6.5, cols. 504–5; cf. 3.19.4, cols. 615–16. *In psalmos* 32:229, 372; *In Iob* 33:544.
229. *Institutio* 3.25.2, cols. 729–30.

Harvey, by Hercules! The Hero of the Blood's Circulation

The virile oath *mehercule* exceeded the discipline of William Harvey's masterful *Exercitatio anatomica de motu cordis et sanguinis in animalibus* (An anatomical exercise on the movement of the heart and the blood in animals) (1628).[1] Medical texts were not in the habit of swearing. Where they did swear—by Hercules, and by Zeus or Jove—marked the arguments Harvey swore in his ambition to alter the history of medicine from tradition to science.[2] Although the exclamation has been credited as "colorful prose,"[3] most English translations have avoided its literal sense. Renditions have been "by my troth," "in fact," "in faith," "damn it," "damme," and "in God's truth."[4] A single editorial revision told the truth, "by Hercules."[5] The translator commissioned by the Royal College of Physicians for the tercentenary version rehearsed previous efforts, then explained his own expletive. "It took me two hours to think of 'damme' as a sufficiently dramatic translation of 'mehercule,' but the *Oxford English Dictionary* then assured me that in the form 'Damn me' it was in use as early as 1645, and I felt that the two hours had been well spent in getting the effect that Harvey had, to my mind, intended by his use of a particular Latin word."[6] Yet, the synonym for *mehercule* in good Renaissance English was "by George," for the dragon slayer and national saint.[7] Harvey chose

This chapter was originally published in *Medical History* 57 (2013): 6–27.

M. O'R. Boyle, *Cultural Anatomies of the Heart in Aristotle, Augustine, Aquinas, Calvin, and Harvey*,
https://doi.org/10.1007/978-3-319-93653-6_5

the anachronism deliberately. His rhetoric displayed his personal character and professional end, as he brilliantly but subversively assumed the persona of Hercules to embody his own labors.

Oaths

Hercules was the classical superman, and masculine swearing by him was commonplace in classical culture. Among texts Harvey studied, Quintilian's *Institutio oratoria*, which governed the English curriculum, swore "by Hercules."[8] Cicero's oratory, which Harvey read in the sixth form at the King's School, Canterbury,[9] swore mightily "by Hercules," from familiar letters to forensic practice.[10] Harvey learned to speak Latin from Terence's comedies, which boasted ninety-six oaths *hercle* (*mehercle* twice). A dozen were in *Adelphi*, the play Harvey soliloquized to the College of Physicians, London, about his role as the discoverer of the blood's circulation.[11] Harvey's *mehercule* was not color but argument. Renaissance rhetoric invented arguments by imitation, best by emulation, which transcended the style of masters by cultivating a native gift.[12] Writers of genius reprised traditional roles to discover and display their own identity.[13] As an emulator, Harvey echoed but parodied oaths in medical masterpieces to assert his own anatomical convictions. Despite his characterization as "very Cholerique" and "hott-headed," his *mehercule* was not verbal swashbuckling like his youthful fashion of wearing and drawing a dagger.[14] Although choler was blamed for swearing,[15] Harvey's oath was not a temperamental outburst. An oath was often reinforced by *certe* or *certo*, "for certain," as a marker of truth.[16] The oath *mehercule* stated truth and elicited assent, as in Plautus's line, "By Hercules, you really speak correctly, and I agree with you."[17] On the London stage, a soldier in William Shakespeare's *Antony and Cleopatra* freely swore "by Hercules, I think I am i' th' right."[18]

Harvey's serious purpose in swearing to his colleagues "by Hercules" was signaled by his prior oath, *Deus bone*. All translations have rendered it faithfully as "Good God!" although none has commented on its oddity in an anatomy book.[19] The excuse of "exasperation"[20] diminished its import. Swearing was common enough in Harvey's society. It encompassed "the calling to witness of something, divine or otherwise, to seal vows of allegiance and promises of love or to attest the truth of a statement; and the inclusion of a similar phrase in a more exclamatory fashion to add emphasis to one's speech."[21] Swearing was a political necessity to secure civil authority and order, especially once Henry VIII's breach with Rome necessitated the Oath of Succession. That reform then required of subjects successive oaths of loyalty to the monarchy, disloyalty to the papacy and other foreign powers.[22] The parliamentary Act of Supremacy would require the oath of fealty thus: "I, *name*, do utterly testify and declare in my conscience ... So help me God and by the contents of this Book."[23] Although in Harvey's days at the University of Cambridge that oath was not enforced among candidates for the bachelor of arts,[24] later in 1605 he was obliged to take the prescribed Oath of Allegiance to James I. The law specified "all doctors of physick, and all others who practice physick, that now are or hereafter shall be admitted into the College of Physicians in London."[25] Harvey did so "trewly and sincerely acknowledge, professe, testifie, and declare in my conscience before God and the world ... And I do make this Recognition and acknowledgment heartily, willingly, and trewly, vpon the trew faith of a Christian. So helpe me GOD."[26] He was further sworn to the king in 1618 as his physician extraordinary.[27] That oath was taken on the King James Bible, with its freshly Englished commandment, "Thou shalt not take the name of the Lord thy God in vain; for the Lord will not hold him guiltless that taketh his name in vain."[28]

Yet, oaths were also sworn liberally and loosely in his contemporary England not only by tinkers but also by gentlemen and those who aspired to their class by parroting their language. The unexpurgated plays of Shakespeare, whose death in April 1616 coincided with Harvey's inaugural Lumleian lecture in anatomy, were explosive with swearing and forswearing, grave and glib. Those dramatic oaths authenticated characters and moved plots. Since pagan oaths in drama were not subject to the censor, the allowance sharply increased resort to pagan deities. Moralists railed in pulpit and press against a populace who swore habitually, from sacred truths "by God" to silly things "by the mousefoot." A canonical sin, subject to punishment in the ecclesiastical courts, profanity became in 1606 under Puritan pressure a statutory crime in public performances and printed texts, subject to a fine of ten pounds. The law was extended in 1623 to general usage, subject to a shilling fine for the relief of the poor or three hours in the stocks. In 1627 the Act was ratified by the parliament of Charles I,[29] to which king Harvey a year later dedicated *De motu cordis et sanguinis.*[30] Harvey was liable to criminal charge for his oath *Deus bone.* Because his book was published abroad, by William Fitzer in Frankfurt, it escaped scrutiny by the newly appointed archbishop of London, William Laud, who licensed medical books and also excised or softened swearing in print.[31] Nor did its dedicatee, Charles I, censure Harvey for his oath

Deus bone, despite the king's pious reputation. Charles was esteemed "so severe an exactor of gravity and reverence in all mention of religion ... that he could never endure any light or profane word."[32] His orthodoxy would be manifest when he imposed the *et cetera oath*, binding all doctors of physick to the established doctrine, discipline, and government of the Church of England.[33] If Charles read Harvey's proem, the oath *Deus bone* did not deter the king from promoting him in 1631 his physician-in-ordinary.[34] Although Harvey's *mehercule* may have been negligible, its first English translation in 1653, "by my troth," was strictly unlawful. In 1635 letters patent established a public department to enforce in each parish the laws against swearing. In 1640 the parliamentary Acts of 1623 and 1627 against swearing were also ratified. On the records, one Thomas Buttand was punished for uttering "on my troth,"[35] so that Harvey's written "by my troth" was spared notably.

The president of the College of Physicians, John Argent, to whom Harvey also dedicated his volume,[36] could not have missed the oaths. Although the College was not legally empowered to censor books, it adjudicated not only malpractice but also comportment.[37] Improper language was denounced and punished. Among its prosecutions of surgeons that Harvey attended were hearings against John Lumkin, who was imprisoned with a fine of twenty pounds for his "abusive language, which was not to be borne," and William Kellet for his "vile language," as "contrary to the ordinance and good government of this house."[38] English physicians commonly swore lawfully, however. They took the ancient Hippocratic *Oath* "by Apollo, Physician, by Asclepius, by Health, by Panacea and by all the gods and goddesses, making them my witnesses."[39] Their College, of which Harvey was a fellow, was a sworn society. Its statutes required an applicant's oath of his nationality, his collegiate duties such as the aid or avoidance of fellows, certain Hippocratic ethics, attendance at anatomical demonstrations, and the reading of stipulated works of Galen. Its statutes also required oaths upon the admission of an officer, which Harvey became, and the passage of statutes, whose revision he assisted. Breach of oath was subject to a fine under penalty of expulsion. The College even summoned under seal the excluded—apothecaries, surgeons, druggists, nurses, and servants of patients—to testify to its president and censors under oath. It had petitioned the king for the right to administer oaths to external witnesses "lest Medicine perish."[40] Its annals recorded Harvey in 1604 taking the oath as a licentiate "according to the form in the Statutes."[41] His further solemn charge in 1609 as physician to the sick poor at St. Bartholomew's Hospital, London, twice enjoined the performance of his duties "in God his most holly name."[42]

Swearing vainly by God's name was sin against the third biblical commandment, which he learned as a requirement for his admission as a boy to the King's School.[43] In his *Prelectiones anatomie universalis* (Lectures of the whole anatomy), Harvey repudiated swearing even by one's head. Regarding the philosophical dispute about the seat of the soul, he considered the head the perfect part whereby humans excelled and dominated. "Whence the head is the most precious member, and to swear by the head and waste the holy is sinful."[44] As the manual that set the style for English behavior, Henry Peacham's *The Compleat Gentleman*, advised, "But aboue all, in your talke and discourse haue a care euer to speake the truth, remembering that there is nothing that can more preiudice your esteeme then to be lauish-tongued in speaking that which is false." He noted that Plato allowed only physicians to lie, and that only for the comfort of the sick.[45] Harvey's oaths in *De motu cordis et sanguinis* swore not vainly but truthfully, not blasphemously but legitimately.

Profanity, the invocation of God or gods to witness the truth, was an ancient type of judicial oath administered by social bodies for protection from lying individuals.[46] Not all Protestant reformers were as incensed about swearing as the Puritans and Anabaptists. Calvin, whose theology dominated the Church of England,[47] decried its promiscuity but retained its legitimacy. As a lawyer turned theologian, he justified not only public oaths, such as swearing fealty to a monarch, but also private oaths taken "soberly, solemnly, reverently in necessary matters."[48] His magisterial *Institutio christianae religionis* argued a case that fit Harvey's contested situation in the College as its lecturer in anatomy. Some fellows, Harvey complained in his proem, falsely accused him of a breach of faith with medical precepts.[49] Fellows were sworn against "speaking ill of fellows," and accusing a colleague was subject to fine for the first offense, fine and expulsion for the second. Harvey had twice served as a censor, so he knew the statutes permitted the investigation of slander.[50] He reported a consensus that he was slandering Galen's medicine, thus Galen's faithful. That accusation provoked his self-defense sealed by those oaths. As justification, Calvin had reasoned from Scripture that "if it is lawful in grave and serious matters for individuals to call God to judge between them, much more so is it lawful to call him as a witness." Suppose, he wrote, your brother accuses you of perfidy, and in charitable duty you strive to clear yourself, but no reasoning satisfies him. Then, Calvin argued, "if public opinion of you becomes divided because of his obstinate ill-will, without offense you may call forth the judgment of God so that in time your innocence may be manifest." The underside of the invocation of divine authority involved a curse upon false or frivolous words. As Calvin explained, "We cannot avail ourselves of God as the witness of our speech without imprecating him as the punisher of our perjury, should we deceive."[51] Harvey's swearing was asseveration, or solemn affirmation, entailing adjuration, a potential curse upon himself for falsehood.[52] The Hippocratic *Oath* ended in such an imprecation on its violation by transgression or perjury. "Now if I carry out this oath, and break it not, may I gain for ever reputation among all men for my life and for my art; but if I transgress it and forswear myself, may the opposite befall me."[53]

Both of Harvey's oaths occurred in his proem, the rhetorical statement of purpose.[54] Both oaths strategically concluded paragraphs condemning the Galenic cardiovascular tradition, to which the College adhered. Harvey swore in order to reject Galen's fundamental belief in the universal power of physical attraction. Galen swore *Dia*, "by Zeus!," precisely to explain the attractive faculties of the heart and the arteries in the blood flow.[55] Harvey invoked *Deus bone* to witness to the falsehood of Galen's system of the exchange of air and spiritous blood between the lungs and heart. "Good God! How do the tricuspid clacks impede the egress of air and not of blood?"[56] By swearing his objection to Galen's physiology, Harvey elevated his own from reasonable thought to a matter of conscience. His charge to care for the patients at St. Bartholomew's Hospital "in God's most holly name" became a mission in the College of Physicians to correct the principal error of traditional medicine. For, that is how Harvey regarded the heart, as the *principium*, analogous to the king as the *principium*.[57]

His pagan oath, *mehercule*, extended his repudiation of the cardiovascular tradition. Harvey's *mehercule* sworn in his *prooemium* paralleled Celsus's *Hercules* sworn in his *prooemium* of the basic Roman medical text.[58] With better latinity, Harvey invoked Hercules in the vocative case and with the preferred Ciceronian usage, *mehercule*.[59] Celsus's review of the medical schools swore, "By Hercules, the ancient doctors were not ignorant of that." He referred to their diagnostic consideration of the commonalities of diseases, following upon an example of professional interest to Harvey. "Those who take charge of large hospitals, because they cannot pay full attention to individuals, resort to these common characteristics."[60] With his own oath, Harvey dismissed Celsus's assertion of ancient medical knowledge, with "by Hercules," the ancients were an ignorant lot! The insinuation, couched in an oath, was thematic of Harvey's proem, which confronted tradition with innovation. The quarrel of the ancients and the moderns, which characterized the Renaissance revival of classical learning against medieval "barbarism," was transferred from literature to medicine—but with reverse valuation. A humanist recourse to correct texts, rather than demonstrated anatomies, Harvey believed had ultimately deterred the advancement of medical knowledge. Harvey's modern method intended to undo classical opinion. But he ingeniously employed classical texts to justify that subversion.

Labors

Harvey impersonated the demigod Hercules by whom he swore the truth. Herakles/Hercules was the outsized hero of classical legends whose mutable literary fortune elastically encompassed the tragic and comic, virtuous and vitious, intellectual and physical.[61] His huge character was imitated by protagonists of Marlowe and Shakespeare who strutted the London stages for edification and entertainment.[62] Of Hercules's literary successors, Vergil's epic hero, Aeneas, was the outstanding role model for Harvey as royal physician and collegiate anatomist. Vergil intended his Herculean heroism to deify the emperor Caesar Augustus for his political labors. Through the influence of euhemerism, which transfigured the classical pantheon into men elevated to gods by their devout descendants, the ideology strongly impressed Renaissance culture.[63] Harvey's epistle dedicatory to his patron Charles I was not simple courtesy. Hercules was since antiquity a model of the ethical ideal king.[64] European royal households claimed descent from Hercules for a legitimacy and authority derived from his essential virtue of fortitude in intense suffering against all odds.[65] Henry, Prince of Wales, the eldest son of James I, was invested with the persona of Hercules from his Scottish Presbyterian baptism. The babe was displayed on a bed of state decorated with a tapestry embroidered with the labors of Hercules, although his beasts to slay were Rome and Spain.[66] At his death, his successor to the throne, Charles I, was celebrated by his courtiers as a new Caesar Augustus, a type of Hercules,[67] as Vergil's *Aeneid* had poetized. Harvey's comparison of the heart, prince of the body, to Charles, prince of England,[68] implied that whatever honor accrued to his anatomical labors would attach to the king. It suggested that *De motu cordis et sanguinis* would earn Charles greater fame than the books by royal physicians Harvey would best: Galen, his translators Thomas Linacre and John Caius, and Andreas Vesalius. In the end his patron suffered regicide, for which the *Book of Common Prayer* atoned. Its office for the fast day of Charles I, martyr, extolled his virtues. But his models were the biblical kings Josiah and David, and Christ in his divine passion[69]—not Hercules in his heroic endurance.

Vergil's Aeneas as a second Hercules was an exemplar of filial piety, memorable in the flight from Troy, in which he carried his father on his shoulders.[70] Harvey in his flight from tradition was not so pious. He paid what respects he could to his medical fathers; ultimately he unloaded them from his back. But that decision also imitated Hercules, who departed egregiously from social morality and its conventions.[71] Hercules did not observe the classical ethic that expected everyone to maintain their stations in life and respect their superiors.[72] Moreover, for his "madness" in slaying his children, Hercules was diagnosed, since the medieval Pseudo-Aristotelian *Problemata*, as a melancholic. He became an Elizabethan type of temporary insanity as a frenzied actor or mad hero. Through Platonist spiritualization of melancholy, he was associated with the ecstatic inspiration of geniuses.[73] As a Hercules, Harvey could thus be acknowledged as a healing physician or mistaken for a delirious patient.

Harvey learned the Herculean legends as a boy from Erasmus's *De duplici copia verborum et rerum*, the statutory textbook for Latin rhetoric at the King's School. To instruct a full and fluent style, it advanced Hercules variously, as an example of an epithet from deeds, "Hercules, reducer of monsters"; of variety by substitution, "the man from Tiryns"; of similitude, "a second Hercules"; of prosperity, the adage "with Hercules' blessing"; of fictional example, "the story of Hercules fighting the twin-horned Achelous." The similitude "a second Hercules" became Harvey's ambition. He learned under "fictional examples" that "the labours of Hercules tell us that immortal renown is won by effort and by helping others."[74] From Horace, also required reading, he learned that Hercules was "the man tenacious of his purpose in a righteous cause," who earned the "starry citadels" with the gods.[75] Erasmus's *De ratione studii*, frequently printed with *De copia*, advised schoolmasters that topics for boys should not be empty and dull but instructive and delightful. It proposed the mythological example that "Hercules won immortality for himself by vanquishing monsters." Since it recommended for boys exercises based on fables,[76] Harvey may well have practiced his Latin letters on storied Hercules.

The Renaissance exemplar of "the labors of Hercules" was Erasmus himself. His *Adagia*, the massive collection of classical sayings that included it, impressed English culture, for the most striking feature of early seventeenth-century English speech was its proverbial habit.[77] Erasmus's volume was researched in England, dedicated to an English patron, and concluded with a letter to Prince Henry (King Henry VIII) and a poem in praise of England. Enthusiastic readers there culled it for their compositions, and also translated, annotated, and incorporated it into native dictionaries. Copies sold to individuals, from penurious scholars to wealthy collectors, who bequeathed them in their wills. College libraries shelved it.[78] *Adagia* was in the basic curriculum.[79] In his essay on "the labors of Hercules" Erasmus explained one meaning as "continuous and very great exertions, and as such as demand Herculean strength." He promoted his own literature as surpassing the norm with "much more than Herculean labors." That definition coincided with Harvey's medical achievement. The other meaning of the adage fit Harvey's situation before his critics. Erasmus identified it as "tasks of the kind that bring very great blessings to other people, but almost no return to the man who undertakes them, except a little reputation and a great deal of ill-will." Harvey attributed his detractors' motive to *invidia*, "envy," a passion he diagnosed as pathogenic because it altered the heart and caused diseases.[80] His blame echoed Erasmus's repeated equation of Hercules's labors with the provocation of *invidia*, "envy." Erasmus quoted Horace's epistle to Caesar Augustus on how Hercules learned at the end of his labors that the last monster to be vanquished was envy.[81] In English experience, George Chapman compared his translation of the twice twelve books of Homer's *Odyssey* to the twelve labors of Hercules and complained of the envy the deed invited.[82]

But Harvey was not intent on the translation of Greek classics, even their medical texts. His oath "by Hercules" desired to achieve by his anatomical demonstrations the immortality that mortal won for his heroic deeds. From his folkloric origin as a strong boy, Hercules embodied perfect *physis*,[83] so that a doctor, practitioner of "physicke," was a credible successor. A Herculean role for a physician was classical, and a doctor could boast of being a descendent. In antiquity Hercules could be vowed a tithe of property in exchange for a healing, or erected a fountain in honor of the god more effective than doctors.[84] The emperor and philosopher Marcus Aurelius swore his health "by Hercules."[85] Although mythology did not associate Hercules and medicine, cult evidenced his role in healing. The association may have derived from his role as *alexikakos*, "protector against evil," through the hard labors that witnessed to his endurance in suffering. Hercules became paralleled or conflated with Aesculapius, the god of medicine. A Herculean motif was prolific on classical surgical instruments: probe, retractor, grip, scalpel, handle, elevator, curette, strigil, and holder. Among the naturalistic designs, probably the most popular was the bark and knot of a tree limb or trunk to symbolize Hercules's club. Knife hafts were realistically shaped like Hercules's bust in his lion skin, and a retractor's finial copied the head of the Nemean lion he slew. A Herculean motif was most frequent on instruments that caused or alleviated pain. When surgeons gripped their Herculean knives, they sought to imitate his bold righting of wrong, while patients were reminded of his patience in danger and pain.[86]

A medical Hercules was prototypical in Hippocrates, father of medicine, who traced his ancestors to him. Pliny's *Naturalis historia* recorded Hippocrates's prescience of a "plague" and dispatch of his disciples to attend it, "for which service Greece voted him the honours that it gave to Hercules," initiation in the Eleusinian mysteries at public expense and Athenian citizenship. Hippocrates's legendary remedy for that Athenian "plague" was a conflagration in the public streets to counteract airborne miasma.[87] Harvey knew the report because in imitation therapeutic bonfires against plague were a regular practice in London's streets.[88] Horror and duty weighed on him and those collegiate fellows summoned in 1625 by the lord mayor to confer with the aldermen toward a solution.[89] While Harvey's aspiration to be "a second Hercules" had an ancient predecessor in Hippocrates, it had a Renaissance pretender in England in Thomas Lupset. The preface to Galen's commentaries on Hippocrates in the first edition praised Linacre's translations and credited Lupset for editorial assistance. "Nor should you belittle in this service Lupset, who exerted himself with every sinew," laboring like "a second Hercules."[90] In the mold of the humanist and physician, Lupset was a graduate of St. Paul's School in London, of Cambridge University, and of Padua in medicine. He supervised the printing of Linacre's translations of Galen's *De sanitate tuenda* and *De methodo medendi*.[91] The target for a Herculean club was overkill for English medicine, however, which was not very influenced by humanist methods.[92]

Harvey sought to usurp the epithet "a second Hercules" by reforming humanist dependence on ancient texts as authoritative medicine. He would correct its misapplication, exemplified by that Greek edition of

Galen and abetted by erudite translations of its errors.[93] In Harvey's judgment, scholarship was dedicated to texts, rather than to medicine. Although its publications might promote understanding of traditional medicine, they certainly prolonged its errors. Humanism was not a subject but a method. Derived from the trivium, it favored, against the hegemony of scholastic logic, the application of classical grammar to texts and classical rhetoric to arguments.[94] Humanism as such did not endorse Galenism as medicine but as ancient literate argumentation superior to medieval barbaric reasoning. When Erasmus castigated the Aldine text of Galen as "lies and sacrilege,"[95] he did not complain about its medicine but its manuscripts. As his commentary on the adage "the labors of Hercules" explained, his humanist criticism concerned "the prodigious corruption of the texts, which has acquired such a hold upon all our copies of both Latin and Greek authors that, whatever you touch in hopes of quoting it, you hardly ever have the good fortune not to stumble over some obvious error or suspect one below the surface." Erasmus's solution was to encourage the acquisition and collation of more copies,[96] not the requisition and dissection of more cadavers. His own translations of Galen were not of medical but propaedeutic works, arts. Renaissance humanism commendably restored ancient medical texts that had been garbled in medieval transmission. But, for medicine, whose progress needed anatomical demonstrations above literary models, humanism was marginal. What medicine required, Harvey believed, were not critical editions of ancient texts but clinical investigations by modern anatomists. As he declared, he set his mind to "observation," to invent the method for his own book "from many dissections … through autopsies and not through the books and the writings of others."[97] He accomplished that task as a practiced anatomist who could argue classical literature subversively against the propriety of humanism for medicine.[98]

Beyond the humanist Lupset as "a second Hercules," there was another possessor of the title Harvey coveted. That was the celebrated anatomist Vesalius, whose *De humani corporis fabrica* Harvey cited most often among modern authors. Vesalius's anatomy professor at the University of Paris, Johann Guinther of Andernach, praised his young prosector in his own anatomical textbook. Guinther swore about the discovery of the spermatic vessels by "Andreas Vesalius, the son of the emperor's apothecary, by Hercules (*me hercules*) a youth of great promise, and extraordinary knowledge of medicine, also accomplished in both languages, and very dexterous in dissecting bodies." That boast falsely credited his student with a discovery that other anatomists had published—Niccolò Massa in the same year, but previously Mondino de' Luzzi and Galen.[99] Vesalius published an unauthorized emended edition of his professor's textbook to correct its typographical errors; but he did not correct that falsehood about himself.[100] Vesalius knew the import of swearing the truth "by Hercules." In his *Examen* he confessed that when he rent the substance of the veins to find the fibers, as Galen taught, and dissected the matter raw and cooked, "by Hercules" (*mehercule*) the fibers were imaginary.[101] Since Guinther's text was the first manual for anatomy students, Harvey knew its double and duplicit swearing "by Hercules" on behalf of an alleged discovery. To restore integrity to anatomy, Harvey imitated Vesalius's oath "by Hercules" on Galen's imaginary venous fibers with his own oath on Galen's imaginary porous septum.

Harvey confessed that when he applied himself to understand the movement of the heart and the blood, from the very beginning he found it "a matter quite arduous and immediately full of difficulties." He only extricated himself "from this labyrinth" by climbing up its steep slope and out. "I got it," he finally wrote.[102] How? by Hercules. Harvey's oath *mehercule* not only invoked that immortalized mortal to witness to the truth of his own arduous and difficult labors, it also intimated how Harvey invented about the blood's circulation. Or, rather, it indicated how Harvey chose to publicize that; for, invention and disposition were distinct parts of argumentation.[103] Hercules was legendarily ordered to atone for his mad slaying of his children by performing twelve mighty labors. Harvey identified his own anatomies as "labors,"[104] and he imitated Hercules's labor of the cleansing of the Augean stable. In classical tales with various details, Augeas, king of Elis on the Greek mainland, kept a stable of three thousand oxen that had not been cleaned in thirty years. Hercules was assigned the formidable task of removing their dung in a single day. He mucked the Augean stable not with shovel, mop, or broom but by his wits. Noticing across from the stable the convenience of an adjacent river (or two), Hercules engineered a solution. He dug a hole through the foundations of the cattle pen, then an outlet on its other side. Then he diverted the course of the river by damming it. Thus rerouted, the river surged, then gushed through the breech in the stable, washing out its filth with powerful water pressure. A bonus was that the ordure flushed through the stable fertilized the fields beyond for a bumper crop. But Hercules was refused the promised payment for his labor, a tithe of the oxen.[105]

Hercules's feat has been reduced in modern translation to inventing "the world's first sewage drain."[106] But in Renaissance literature the cleansing of the Augean stable had rich cultural interpretations. Coluccio Salutati's *De laboris Herculis* reported the invention of dung as fertilizer but allegorized that labor for cleansing the filth of vices.[107] Hercules's cleaning job was solemnized by a comparison with Christ's death on the cross, which rid the world of the detestable stench of sin.[108] Erasmus explained the adage "to cleanse the stable of Augeas" as "a proverbial allegory, used of a person or thing that is filthy beyond measure."[109] In Harvey's time, Leone Allacci, a scriptor at the Vatican Library, exposed an Etruscan forgery as "a new Augean stable, full of foul odors and outrage ... while others gather flowers from the manure, I collect, albeit necessarily, manure."[110] It was the secular interpretation of Hercules's labor, about the reform of learning, that Harvey appropriated. Erasmus importantly revived the cleansing of the Augean stable in his youthful manifesto, *Antibarbari*, to promote the humanist purge of scholasticism, whose "barbarism" was "filth." False friars who regarded themselves as demigods were attacking literary studies. Yet, "if anyone dares to divulge any of their secrets [vices] and disturb the Augean stable, they announce that he is in danger of destruction from an irate Francis, or Dominic, or Elijah, so help me!" The interlocuter Jacob Batt, a local stand-in for Erasmus's universal labors, recalled how he undertook "the labors of Hercules" as a schoolmaster. "What an Augean stable I found there. Ye gods! what nonsense, what inanities, what mockery, what barbarism, what thorns and brambles, what dregs had been forced upon the unhappy schoolboys by those before me who had taught them to know nothing." His pedagogical reforms enraged the citizenry, who accused him of immorality, of heresy, of threatening the end of Christianity and the coming of the Antichrist. He summoned his conversationalists, "You were a witness of that fight; you saw for yourself how I acted Hercules, how many lions and boars and bulls and Stymphalian birds I slew, how many versions of Antaeus or Geryon or Diomedes or Nessus, how I dragged Ceberus out of his den where he was terrifying the pallid shades, and held him up to the sky; you saw how my Greek fire only just managed to wipe out the Lernaean Hydra, fertile with its own deaths, and I rather think that worst of all plagues is still alive and breathing.... Alone I faced all those monsters, but I did not give way; no, I won through, and convinced the saner intelligences, confuted the others with clear reasoning, and held some up to scorn." Batt's companions congratulated him as "Hercules" for his feats and arranged a triumphal procession to honor him. "We will make you a god, so that you will not be a second Hercules, as the proverb says, but Hercules himself; when did he ever deserve so much?"[111]

Cleansings

With his oath *mehercule* Harvey meant to cleanse the Augean stable of medicine as Erasmus had cleansed that of literature. It was not the first application of that Herculean labor to physiology. Jacobus Sylvius upbraided his former student Vesalius whose *De fabrica* dared to slander Galen and by implication him. Sylvius accused Vesalius of violating the Hippocratic *Oath* of respect to teachers. "This slanderer wickedly renounced his oath of allegiance to his master Hippocrates, in which he had promised the greatest gratitude to his teachers and to their adopted children, and, furthermore, that he sought in every way to criticise them falsely, since he hoped that by competing with such teachers for the leadership in anatomy he might some day acquire reknown." Sylvius pondered his task of repudiation, reckoning grammar before physiology. "Everything was so filled with grammatical and other errors, as well as an ignorance of physiology, that it would have been easier to cleanse the Augean stable than to remove even the worst lies from this hodgepodge made up of thefts and bloated with slanders."[112]

Hercules's cleansing of the Augean stable related to Harvey's invention of the blood's circulation exactly where Harvey swore *mehercule*. His rhetorical signal was where it classically belonged, in the *prooemium*, whose synonym was *principium*, his designation for the status of the heart in the body. Harvey wrote, "That opinion is less tolerable that supposes that, since dual matter (airy and bloody) is necessary for forming the vital spirits, it contends that the blood sweats across through the invisible porosities in the drudge of the heart from the right into the left ventricle, and air is drawn through the great vessel, the arterial vein; and accordingly that in the *septum* of the heart there are very many porosities for admitting the blood. But, *by Hercules*, there are no porosities, nor can they be demonstrated."[113] The argument repudiated Galen's notion that the cardiac ventricles were separated by a porous septum. Through its minute invisible holes the finest particles of the venous blood supposedly passed from the right to the left ventricle for purification. Renaissance anatomists agreed and marveled, or wondered and doubted. Matteo Realdo Columbo, Vesalius's successor at Padua, argued in *De re ana-*

tomica against the porosity of the interventricular septum. He rerouted the blood flow from the right ventricle of the heart through the pulmonary artery to the lungs, then from there to the left ventricle of the heart through the pulmonary vein.[114] Harvey's proem, which swore by Hercules against Galen's porous septum, acknowledged Columbo's "opinion."[115] Neither acknowledged Miguel Servet's speculation about his theology of the soul in *Restitutio christianismi* against the septum and for a transfer of the blood through the lungs.[116]

A medical *septum* was a membranous separator.[117] An ordinary *septum* was any "fence, enclosure, wall," but particularly a "cattle-fold," or enclosure for livestock.[118] Harvey knew both senses, for the word appeared in Vergil's *Eclogues*, which he cited on the title page of his *Prelectiones.*[119] In that poem, the enslaved shepherd Tityus abandoned the "folds (*saepta*)" to travel to Rome as a freedman.[120] For Harvey, the cardiac septum was like a cattle-fold that he would cleanse by abandoning it. Since in Galen's anatomy the septum admitted the blood through it, Harvey had to divert that flow. To cleanse the Augean stable of medicine Harvey dammed the septum, which allegedly sweat, seeped, or trickled blood across the heart through undemonstrated porosities. By vivisection and dissection, he had been unable to observe any such openings. As he emphatically swore, "*by Hercules*, there are no porosities, nor can they be demonstrated." He concluded his refutation of the septum with the necessity of opening an alternative route. Diverting the blood flow from the septum, he channeled it continuously through the arteries and veins. Harvey argued that circulation was consistent with the pulse of the heart and arteries, with the transfusion of blood from the veins into the arteries, and with their distribution of the blood throughout the body.[121] Harvey knew about the use of rivers to clean stables on an explicitly circular model, for Padua, where he studied medicine, was a round city encircled by rivers.[122] Fynes Moryson, an English tourist in 1591–1595, recorded the course of Alpine rivers. "These Rivers enter the City, and with divers channels drive many mils, compasse the wals, and not onely make the fields fertile, but serve to carry all commodities (abounding here) from hence to Venice, and to bring from thence such things as they want." He ended with a Herculean task "and besides doe cleanse all filth of the stables and privies."[123]

Harvey's circulation further imitated Hercules's successful dependence on the rushing force of the water flow to cleanse the Augean stable. That efficacy Harvey applied *de copia sanguinis.*[124] Although honored as his "quantitative argument,"[125] Harvey wrote *quantitas*, paralleled with *pro-*

portio, only once. He emphasized *copia* in two chapter titles and repeated it twenty-one times more.[126] The primary lexical meaning of *copia* is not quantity but "abundant power." Harvey borrowed his chapter titles, *De copia sanguinis*, from the textbook of exercises in style at the King's School. As its royal charter prescribed, "Finally, in the sixth form they essay those formulas *De copia verborum et rerum* written by Erasmus and learn to vary rhetoric in numerous ways, so that they might thus attain a faculty in the Latin language (as much as is sufficient for boys)."[127] Human speech, Erasmus began suggestively, "is a magnificent and impressive thing when it surges along like a golden river, with thoughts and words pouring out in rich abundance." He advised recourse to literary passages where "the spring of eloquence seems to bubble up particularly richly." That concept of *copia*, "powers of expression," did not denote quantity, although some amount was necessarily connoted. It was a "godlike power of speech" distinct from the "excessive verbosity" of "mere glibness." Students were warned against sheer quantity—not to "pile up a meaningless heap of words and expressions without any discrimination." Although *copia* drew from a store of language and material, that treasury was to be commanded judiciously and elegantly for appropriateness to the subject and audience. In fact, brevity might be required, but again not in consideration of quantity—not "to say as little as possible, but to say the best things as briefly as possible." Erasmus forbade "excessively long digressions at inappropriate points." He counseled "due account of order and arrangement lest a mass of unorganised material throw the whole speech into confusion and disorder." In studying *De copia*, Harvey learned not how to count but how to develop resourcefulness and judgment.[128] What he paralleled with *De copia sanguinis* was "blood supply." Although he observed the blood supply as plentiful (*quanta copia*), his emphasis was not its measured amount but its ready availability. Harvey's *copia* intended function, "blood supply" for a useful and appropriate purpose. In traditional physiology, blood was concocted in and consumed by the body in a terminal process that required renewal. Harvey's observation, during numerous dissections, of the blood spurting liberally suggested another possibility. "I, *I* began to think to myself whether it might have some movement as if in a circle, which afterwards I found out to be true."[129]

From that intuition to its proof, Hercules's feat of hydraulic engineering to cleanse the Augean stable was a model for the rhetorical disposition of Harvey's labors. He argued for the circulation of the blood, as his proem with the Herculean oath stated, by "ocular demonstrations" and "reasoned arguments,"[130] that is by comparative anatomy and natural philosophy. But he presented his achievement to his colleagues by adopting the Herculean medical persona, and that was a rhetorical imitation that personified those labors. His was not a generic role, as in Wolfgang Höfer's *Hercules medicus*, which will illustrate and praise every physician as clubbing the Hydra of diseases.[131] Harvey's particular labor was the cleansing of the Augean stables, an excellent comparison. The damming of the water/blood flow for diversion into a new channel and the efficacy of its powerful supply accomplished the deed for the first and second Hercules.

Initially tossing to and fro mentally to understand cardiac movement, Harvey compared himself to Aristotle, for whom "the motion of heart was like the flux and reflux of Euripus."[132] Harvey repeated the anecdote from the physician André DuLaurens,[133] whose late scholasticism relied on the authority and reasoning that perpetuated such ignorance.[134] But the medical subtext was Galen's *De usu partium*, which rejected the comparison of the blood flow to the tidal flow in the straits of Euripus.[135] The classical source for the anecdote about Aristotle was Procopius's history, which recorded how the phenomena of currents in straits "appear to be susceptible of no explanation, nor has anyone ever shewn himself able to account for them." As Procopius foretold the issue of authority that confronted Harvey in the College, "I am aware that as a general thing all men, if they first discover an ancient argument, are no longer willing to devote themselves to the labour involved in the search for truth nor to learn instead some later theory about the matter in hand, but the more ancient view always seems to them sound and worthy of honour, while contemporary opinions are considered negligible and are classed as absurd." Euripus was a vexing example of a universal problem. "Indeed," Procopius continued, "it was this question which led Aristotle of Stagira, a man prominent among all others to go to Chalcis on Euboea, where he observed the strait which they call Euripus in an effort to discover by careful investigation the physical reason why it is and in what manner it comes about that sometimes the current of the strait flows from the west, but at other times from the east." Sailors navigated by the temporary direction of the current, reversing course as required by the tidal inflow and outflow. "All this the Stagirite observed and pondered for a long time, until he worried himself to death with anxious thought and so reached the term of his life."[136]

The tidal phenomenon of Euripus was well publicized, from Strabo's geography, to Pliny's natural history, to Boethius's philosophy.[137] For contemporary English readers, Nathanael Carpenter's *Geography* recounted the ebb and flow of the sea as "one of the greatest difficulties in Naturall Philosophie: insomuch as Aristotle one of the acutest of Philosophers, is reported to haue stood amazed at the flowing and ebbing of Euripus, and despairing of finding out the cause, at length enforced to cast himself into the River which had before confounded him."[138] Even after Harvey's days at Cambridge, a student's notebook recorded the rhetorical question "Did Aristotle drown himself in Euripus," and his terse reply, "I do not think our Prince was so insane."[139] The anecdote was topical to praise scientific discovery that surpassed traditional natural philosophy. Francesco Stelluti, a member of the Accademia dei Lincei, introduced his colleague Galileo Galilei's *Il Saggiatore* with this praise: "Once an ancient sought the cause,/Near to Chalcis so men tell,/Why Euripus' wave no pause/ Made in ebb and flow as tides now rose and fell;/Had you told him what so well/You have proved of your belief/He would not have plunged into those waves from grief."[140] By citing the topic of Euripus, Harvey pleaded his case for his solution of Aristotle's worry as deserving of fame.

Harvey early learned that tidal phenomenon from Erasmus's resourceful *De copia*. A technique of expressing a superlative varied a noun with a comparative, such as "more restless than Euripus." In his textbook of similes, *Parabolae*, its ebb and flow typified inconsistent persons. His great compilation of *Adagia* repeated the saying "Man's a Euripus" for changeable persons and fortunes.[141] Euripus fit the change of mind Harvey pressed on his colleagues to stabilize his challenged appointment as their lecturer in anatomy. Euripus also presaged his task of Herculean heroism. In Seneca's drama *Hercules Oetaeus* a dying Hercules sacrificed himself to Jupiter on a cliff above the straits of Euripus and there won immortality.[142] Erasmus recalled it in his textbook on letter writing as an example of philosophical mysteries cloaked in legend. The moral of Hercules's immolation above Euripus was that "immortality was only the lot of those who had spent their whole life in honourable labours and unquenchable ardour for virtue, and had waged war tirelessly upon the monstrous apparition of all vices."[143] Thomas Heywood's *The Brazen Age* dramatized it for English audiences: how with his "flesh frying with poyson" Hercules wished to plunge into the cooling straits of Euripus.[144]

Mentally in such straits, Harvey wrote that the problem of the heart's movement "fluctuated" to him with "contrary, various, and confused" indications, like the reversing ebb and flow at Euripus.[145] That echoed his professor at Padua, Girolamo Fabrici d'Aquapendente, on the "perpetual flux and reflux of the blood."[146] Fabrici had cut an incision in a vein to explore its wickets (*ostiola*), "some exceedingly slender tiny membranes in the internal hollow of the veins." He decided that the membranes regulated the blood flow, which he abandoned in the feet.[147] It was at that fork in the anatomical path that Harvey made the topical choice of Hercules. Erasmus's *De copia* had further taught him "Prodicus' invention about Hercules debating whether he should enter on the steep uphill path of virtue, or the downhill path of pleasure." The choice of Hercules between divergent routes ϒ, a Renaissance convention for moral decision,[148] Harvey converted to physiology. Like Hercules at the crossroads, he chose the right, upward path and reasoned that the blood flowed from the foot of the hill/body back upward.

The venous wickets also reminded Fabrici of "knots in the fine shoots of plants." He illustrated his simile by juxtaposed engravings of a ligated arm with the venous wickets and a branch of the herb verbena having similar knots at its sprouts.[149] Botanical knots were granted since antiquity to Hercules as his principal attribute, the knotty club cut from a branch. It figured in Seneca's drama about his immolation unto immortality on the cliff above the straits of Euripus.[150] Hercules's knotty club loomed colossally in Padua. He had already figured in the foundation of that noble city, as Angelo Portenari related in *Della felicita di Padova*, by descendants from a son or companion.[151] The famed Renaissance sculptor and architect Bartolomeo Ammannati carved *Hercules Bestiarius* from eight blocks of local stone and erected the statue in the courtyard of Marco Mantua Benavides, a jurist at the university. Visible through a triumphal arch and above the walls, the colossus stood twenty-five feet high. The surfaces of an octagonal plinth of five feet more depicted in bas relief Hercules's seven feats of slaying beasts and his symbolic apotheosis. On the hero's club Ammannati carved his own name.[152]

The untying of knots, such as the plant-like nodes of Fabrici's venous membranes, was conventional for the solution of riddles or puzzles. Hercules legendarily invented a knot that was difficult or impossible to undo, hence the proverbial "Herculean knot" for an insolvable problem.[153] The Herculean, or square, knot also had a medical application, reported in Pliny's *Naturalis historia* as a special power to heal the wounds it bound. Oribasius's collection of medical texts included Heraklas's description of Greek surgical knots, with instructions on how to tie tight the Hercules knot.[154] Harvey was required in his first year as lecturer at the College to lecture surgeons biweekly on Oribasius's text,[155] so he knew and perhaps demonstrated Hercules's knot. For Harvey, cardiac movement was such a problem that it initially seemed "almost known to God alone."[156] Harvey's resolution of its operation thus elevated him to quasi divine status, like the immortality Hercules won for his deeds. Medical divinization was not arrogant but topical. Erasmus's declamation on medicine rehearsed its cliches, how medical discoverers were revered as gods, like their founder, Asclepiades, who was equated with the honors of Hercules. As Erasmus emphasized, "If immortality is something which is to be desired, then it is achieved as far as is humanly possible by medical research, which can prolong life almost indefinitely."[157] Immortality was more commonly achieved through progeny, however, and Harvey acknowledged that reality in his lecture on the genitals as "the string tied to eternity."[158] By sexual acts of procreation, the species continued, and the individual lived on, as it were, through successive generations. But William Harvey and his wife, Elizabeth, were childless.[159] His only chance for the progeny that ensured immortality was through the popular topic of the book as a child.[160] A colleague Martin Llewelyn will in 1653 preface a translation of Harvey's *Disputations Concerning the Generation of Animals* with a poem praising his books as issues of his brain although his loins had none.[161] The *History of the Worthies of England* will posthumously award Harvey his children by naming his books. "The Doctor, though living a bachelor, may be said to have left three hopeful sons to posterity."[162]

The Hippocratic *Oath* that physicians swore not only aspired to the practicality of a good reputation but also prayed for the glory of an eternal fame.[163] Erasmus's *De copia* taught young Will how to swear for immortality in the best Latin style. For variation, it transformed a declaration into an oath. "'Nothing do I hold more dear or more to be preferred than glory' became 'I'll be damned if I honour anything more than glory.'"[164] The College of Physicians belatedly acknowledged Harvey's "labors" by carving Hercules's epithet "immortal" on his statue.[165] An observer John Collop, M.D. poetized "On Doctor Harvey" and his Herculean labors, praising his cleansing of the Augean stable of medicine:

> *Non datur ultra Hercules* pillars show,
> Beyond a *Hercules* labours thou dost go.
> Sev'n headed *Hydra*, error multiply'd
> Thou need'st no Club, thy knife can soon divide:
> *Augean* filths no work when vy'd with thee,
> Do'st cleanse the Jakes of all antiquite…[166]

Notes

1. William Harvey, *Exercitatio anatomica de motu cordis et sanguinis in animalibus*, facsimile rpt. of Frankfurt: William Fitzer, 1628 (Birmingham, Ala.: Classics of Medicine Library, 1978), p. 18. Translations of Harvey's texts are mine, except as noted.
2. See Marjorie O'Rourke Boyle, "William Harvey's Anatomy Book and Literary Culture," *Medical History* 52 (2008): 73–90; "Reprising Terence's Plot: William Harvey's Soliloquy to the College of Physicians," ibid., pp. 365–86; "Harvey in the Sluice: From Hydraulic Engineering to Human Physiology," *History and Technology* 24 (2008): 1–23.
3. Donald Proctor, "William Harvey (1578–1657): Blood Circulates," in idem, ed., *A History of Breathing Physiology* (New York: Marvel Dekker, 1995), p. 69.
4. Anonymous (trans.), *The Anatomical Exercises of Dr. William Harvey "De motu cordis" 1628: "De circulatione sanguinis" 1652: The First English Text of 1653*, ed. Geoffrey Keynes (London: Nonesuch, 1928), rpt. in part two of Harvey, *De motu cordis et sanguinis*, p. 13; Michael Ryan, trans., "The Anatomical Exercitations of William Harvey M.D.," *London Medical and Surgical Journal* 1 (1832): 591; Robert Willis, trans., *The Works of William Harvey, M. D.* (London: Sydenham Society, 1847), p. 17; Chauncey D. Leake, trans., *Exercitatio anatomica de motu cordis et sanguinis* (Springfield, Ill.: Charles C. Thomas, 1931), p. 21; Kenneth J. Franklin, trans., *Movement of the Heart and Blood in Animals* (Oxford: Blackwell Scientific for the Royal College of Physicians of London, 1957), p. 19; Gweneth Whitteridge, trans., *An Anatomical Disputation Concerning the Movement of the Heart and Blood in Living Creatures by William Harvey* (Oxford: Blackwell Scientific, 1978), p. 20; Emerson T. McMullen, trans., *William Harvey's 'De motu cordis': A New Translation and Latin Edition* (Bethesda, Md.: Academica, 2005), p. 115.
5. Alexander Bowie (London: George Bell and Sons, 1889), p. 18. Cf. *An Anatomical Dissertation upon the Movement of the Heart and Blood in Animals Being a Statement of the Discovery of the Circulation of the Blood; Facsimile Reprint of Frankfurt 1628 with a Translation and Memoir* (Canterbury: G. Moreton, 1894), p. 17. Although no translator is credited, the memoir is signed "B," likely for Bowie. Cf. only "pores" for "porosities."
6. Kenneth J. Franklin, "On Translating Harvey," *Journal of the History of Medicine* 12 (1957): 17–18.
7. For George, see Robert Graves, *"Lars porsena," or the Future of Swearing and Improper Language*, 2nd ed. (London: Kegan Paul, Trench,

Treubner, 1936), pp. 8–9; Ashley Montagu, *The Anatomy of Swearing* (New York: Macmillan, 1967), p. 117.

8. Quintilian, *Institutio oratoria* 1.4.7, 2.5.4, 2.16.12, 6.1.43, 6.3.74, 10.2.3, 12.1.7, 12.6.4. For his supremacy in the English curriculum, see Thomas W. Baldwin, *William Shakespeare's Small Latine and Lesse Greeke*, 2 vols. (Urbana: University of Illinois Press, 1944), 2:197–238.
9. "Cathedrals of the New Foundation, 1541," in Arthur F. Leach, *Educational Charters and Documents, 598 to 1909* (Cambridge: Cambridge University Press, 1911), p. 468.
10. Charleton T. Lewis and Charles Short, *A Latin Dictionary* (Oxford: Clarendon Press, 1969), s.v. "Hercules."
11. "hercle," "mehercle," in *Lexicon Terentianum*, ed. Patrick McGlynn, 2 vols. (London: Blackie and Son, 1963), 1:231–32, 454–55. See also Frank W. Nicholson, "The Use of *hercle* (mehercule), *edepol* (pol), *Ecastor* by Platus and Terence," *Harvard Studies in Classical Philology* 4 (1893). Harvey, *De motu cordis et sanguinis*, p. 21.
12. Thomas M. Greene, *The Light in Troy: Imitation and Discovery in Renaissance Poetry* (New Haven, Conn.: Yale University Press, 1982); George W. Pigman III, "Versions of Imitation in the Renaissance," *Renaissance Quarterly* 33 (1980): 1–32; Jo Ann Della Neva, "Reflecting Lesser Lights: The Imitation of Minor Writers in the Renaissance," *Renaissance Quarterly* 42 (1989): 449–79.
13. See, e.g., Charles Trinkaus, *The Poet as Philosopher: Petrarch and the Formation of Renaissance Consciousness* (New Haven, Conn.: Yale University Press, 1979), pp. 9, 23–24; Stephen Greenblatt, *Renaissance Self-Fashioning: From More to Shakespeare* (Chicago: University of Chicago Press, 1980), pp. 1–9, 256–57.
14. John Aubrey, "Account of William Harvey," Appendix I in Geoffrey Keynes, *The Life of William Harvey* (Oxford: Clarendon, 1966), pp. 434, 435. Cited by Franklin, "Translating Harvey," pp. 17–18.
15. Frances A. Shirley, *Swearing and Perjury in Shakespeare's Plays* (London: George Allen and Unwin, 1979), p. 12.
16. Montagu, *Anatomy of Swearing*, pp. 31–32.
17. Plautus, *Mercator* line 411.
18. Shirley, *Swearing and Perjury*, p. 2; Geoffrey Hughes, *Swearing: A Social History of Foul Language, Oaths, and Profanity in English* (Oxford: Blackwell, 1991), p. 104. Shakespeare, *Antony and Cleopatra* III.vii.67. For that play as Herculean, see Eugene M. Waith, *The Herculean Hero in Marlowe, Chapman, Shakespeare, and Dryden* (London: Chatto and Windus, 1962), pp. 113–41. For pagan oaths in Shakespeare's classical plays, see Shirley, *Swearing and Perjury*, pp. 126–33. For Shakespeare's use of Hercules, see Adrian Poole, review of Charles Martindale and A. B.

Taylor, eds., *Shakespeare and the Classics*, in *The Times Literary Supplement*, 29 July 2005, p. 10.

19. Harvey, *De motu cordis et sanguinis*, p. 6; trans., anonymous, p. 11; trans., Ryan, p. 591; trans., Willis, p. 16; trans., Bowie, p. 16; trans., Leake, p. 18; trans., Franklin, p. 17; trans., Whitteridge, p. 18; trans., McMullen, p. 113.
20. Roger French, *William Harvey's Natural Philosophy* (Cambridge: Cambridge University Press, 1994), p. 99.
21. Shirley, *Swearing and Perjury*, p. xi.
22. William Kerrigan, *Shakespeare's Promises* (Baltimore, Md.: Johns Hopkins University Press, 1999), pp. 28–40.
23. See J. R. Tanner, *Tudor Constitutional Documents, A.D. 1485–1603* (Cambridge: Cambridge University Press, 1922), p. 134.
24. Mark H. Curtis, *Oxford and Cambridge in Transition, 1558–1642: An Essay on Changing Relations between the English Universities and English Society* (Oxford: Clarendon, 1959), pp. 171–72, 194, cf. 51.
25. See George N. Clark, *A History of the Royal College of Physicians of London*, vol. 1 (Oxford: Clarendon for the Royal College of Physicians, 1964), pp. 191–92.
26. James I, *The Political Works of James I*, ed. Charles H. McIlwain (Cambridge, Mass.: Harvard University Press, 1918), pp. 73–74. See also Lori Anne Ferrell, *Government by Polemic: James I, the King's Preachers, and the Rhetorics of Conformity, 1603–1625* (Stanford, Cal.: Stanford University Press, 1998), pp. 22–23, 133–36; Michael C. Questier, *Conversion, Politics, and Religion in England* (Cambridge: Cambridge University Press, 1996), pp. 106–7.
27. Keynes, *Life of Harvey*, p. 137.
28. Ex. 20:7 (AV).
29. See in general Montagu, *Anatomy of Swearing*, pp. 107–72; Hughes, *Swearing*, pp. 55–125. See in particular Shirley, *Swearing and Perjury*, pp. xii, xiii, 4–5, 7–10, and passim; Kerrigan, *Shakespeare's Promises*; Montagu, *Anatomy of Swearing*, pp. 138–53, 157, 159, 163–64; Hughes, *Swearing*, pp. 103, 105, 108; Alvin B. Kernan, *Shakespeare, the King's Playwright: Theater in the Stuart Court, 1603–1613* (New Haven, Conn.: Yale University Press, 1995), pp. 14–16.
30. Harvey, *De motu cordis et sanguinis*, pp. 3–4.
31. See Douglas Bush, *English Literature in the Earlier Seventeenth Century: 1600–1660*, 2nd ed. rev. (Oxford: Clarendon, 1962), pp. 27–28.
32. Edward Hyde, *History of the Rebellion*, ed. W. Dunn Macray, 6 vols. (Oxford: Clarendon, 1888), 4:489, cited by Julian Davies, *The Caroline Captivity of the Church: Charles I and the Remoulding of Anglicanism 1625–1641* (Oxford: Clarendon, 1992), p. 82.

33. Davies, ibid., pp. 82, 275–87. For religion, see also pp. 5–45, 275–402; Kevin Sharpe, *The Personal Rule of Charles I* (New Haven, Conn.: Yale University Press, 1992), pp. 280–82; L. J. Reeve, *Charles I and the Road to Personal Rule* (Cambridge: Cambridge University Press, 1989), pp. 59–98.
34. For the appointment, see Keynes, *Life of Harvey*, p. 279.
35. Montagu, *Anatomy of Swearing*, pp. 166–67, cf. 118; Shirley, *Swearing and Perjury*, p. 9. *Oxford English Dictionary*, 2nd unabridged ed., s.v. "troth."
36. Harvey, *De motu cordis et sanguinis*, pp. 5–9.
37. See Clark, *History of Royal College*, pp. 93, 188–90. For Harvey as a censor, see p. 293; Charles Webster, "William Harvey and the Crisis of Medicine in Jacobean England," in Jerome J. Bylebyl, ed., *William Harvey and His Age: The Professional and Social Context of the Discovery of the Circulation* (Baltimore, Md.: Johns Hopkins University Press, 1979), pp. 7–8.
38. Keynes, *Life of Harvey*, pp. 63, 65.
39. "The Oath," in *Hippocrates*, trans. W. H. S. Jones, 2 vols. (London: William Heinemann, 1923), 1:299. For the swearing, see Heinrich von Staden, "'In a Pure and Holy Way': Personal and Professional Conduct in the Hippocratic *Oath*?," *Journal of the History of Medicine and Allied Sciences* 51 (1996): 409; Charles Lichthenthaeler, *Der Eid des Hippokrates: Ursprung und Bedeutung* (Cologne: Deutscher Ärzte, 1984), pp. 41–48; Leon Edelstein, *The Hippocratic "Oath": Text, Translation, and Interpretation* (Baltimore, Md.: The Johns Hopkins University Press, 1943), p. 50; Steven H. Miles, *The Hippocratic "Oath" and the Ethics of Medicine* (Oxford: Oxford University Press, 2004), pp. 14, 159–64. For Renaissance texts and translations, see Thomas Rütten, "François Tissard and His 1508 Edition of the Hippocratic *Oath*," in *Hippocrates in Context*, ed. Philip J. van der Eijk (Leiden: Brill, 2005), pp. 465–91; Thomas Rütten, "Receptions of the Hippocratic *Oath* in the Renaissance: The Prohibition of Abortion as a Case Study in Reception," trans. Leonie von Reppert-Bismark, *Journal of the History of Medicine and Allied Sciences* 51 (1996): 456–68.
40. Clark, *History of Royal College*, pp. 101, 177, 221, 222, 93, 94, 231, 213.
41. *Annals of the Royal College of Physicians, 1518–1915*, 48 vols. (London: The College, 1518–1915), 2:179a. For the *Statuta vetera*, see Clark, *History of Royal College*, pp. 172–81.
42. See Keynes, *Life of Harvey*, pp. 53–54.
43. "The Refoundation of Canterbury Cathedral and Grammar School, 1541," in *Educational Charters and Documents*, p. 464.

44. William Harvey, *The Anatomical Lectures of William Harvey: "Prelectiones anatomie universalis," "De musculis,"* ed. Gweneth Whitteridge (Edinburgh: E. & S. Livingstone for the Royal College of Physicians, London, 1964), p. 310. Translations mine. For bodily oaths, see also Montagu, *Anatomy of Swearing*, pp. 63, 118; Shirley, *Swearing and Perjury*, p. 20. For the Ciceronian curse "On their own heads be it!," see Erasmus, *De copia verborum ac rerum*, ed. Betty I. Knott, in *Opera omnia* (Amsterdam: North-Holland, 1971–), 1–6:416 (hereafter Amsterdam); trans. Knott, *The Collected Works of Erasmus* (Toronto: University of Toronto Press, 1974–), 24:494 (hearafter *Collected Works of Erasmus*). Cf. Matt. 5:36.
45. Henry Peacham, *The Complete Gentleman, The Truth of Our Times, and The Art of Living*, ed. Virgil B. Heltzel (Ithaca, N.Y.: Cornell University Press for the Folger Shakespeare Library, 1962), pp. 158–59.
46. Montague, *Anatomy of Swearing*, pp. 59–61.
47. G. Lake, "Calvinism and the English Church, 1570–1635," *Past and Present* 114 (1987): 34.
48. John Calvin, *Institutio religionis christianae*, in *Opera quae supersunt omnia*, ed. Eduard Reuss, Eduard Cunitz, and Johann Wilhelm Baum, 59 vols. in 26 (Brunswick: C. A. Schwetschke, 1863–1900), 2:col. 286.
49. Harvey, *De motu cordis et sanguinis*, p. 21.
50. See Clark, *History of Royal College*, pp. 177, 181, 293, 190.
51. Calvin, *Institutio religionis christianae*, cols. 287, 284.
52. For the terms, see Montagu, *Anatomy of Swearing*, p. 105.
53. "Oath," in *Hippocrates*, p. 300; trans., p. 301. See Sanford V. Larkey, "The Hippocratic Oath in Elizabethan England," *Bulletin of the History of Medicine* 4 (1936): 215, 217, 219. See also Miles, *Hippocratic "Oath,"* pp. 163–64.
54. Aristotle, *Rhetorica* 3.14.6 1415a.
55. Galen, *De facultatibus naturalibus*, in *Opera*, ed. Karl Gottlob Kuhn, 22 vols. in 20 (Leipzig: Cnobloch, 1821–31), 2:202. For attraction, see Thomas S. Hall, "Euripus; or, the Ebb and Flow of the Blood," *Journal of the History of Biology* 8 (1975): 324–31. Cf. the oath "*per Iovem*" about the failure of anatomists to dissect humans. Vesalius, *De humani corporis fabrica* (Basel: I. Oporini, 1543), p. 3r.
56. Harvey, *De motu cordis et sanguinis*, p. 16.
57. Ibid., pp. 42, 59. Cf. Harvey, *Prelectiones anatomie universalis*, p. 2, citing Virgil, *Eclogiae* 3.60. See also Erasmus, *De ratione studii*, ed. Jean-Claude Margolin, in *Opera omnia* (Amsterdam), I-2:139–42.
58. Celsus, *De medicina* prooem. 66.
59. Cicero, *Orator* 157.

60. Celsus, *De medicina*, p. 66; first trans. mine; second, trans. W. G. Spencer, 3 vols. (Cambridge, Mass.: Harvard University Press, 1971–79), 1:35.
61. G. Karl Galinsky, *The Herakles Theme: The Adaptations of the Hero in Literature from Homer to the Twentieth Century* (Oxford: Basil Blackwell, 1972).
62. Waith, *Herculean Hero*. Around the turn of the seventeenth century "The Birth of Hercules" was performed at Cambridge. G. C. Moore Smith, *College Plays Performed in the University of Cambridge* (Cambridge: Cambridge University Press, 1923), p. 65.
63. Galinsky, *Herakles Theme*, pp. 35, 128, 129, 131–47. See also pp. 138–39.
64. Ibid., p. 35. See also Marc-René Jung, *Hercule dans la littérature française du XVIe siècle: De l'Hercule courtois à l'Hercule baroque* (Geneva: Droz, 1966), pp. 159–69, 174–77; Ulrich Huttner, *Die politische Rolle der Heraklesgestalt im griechischen Herrschertum* (Stuttgart: Franz Steiner, 1997).
65. Stephen Orgel, "The Example of Hercules," in *Mythographie der frühen Neuzeit: ihre Anwendung in den Künsten*, ed. Walther Killy (Wiesbaden: O. Harrassowitz, 1984), pp. 25, 27.
66. Jerry W. Williamson, *The Myth of the Conqueror, Prince Henry Stuart: A Study in Seventeenth Century Personation* (New York: AMS, 1978), pp. 1–4, 6, 7, 33, 75–76.
67. Robert Wilcher, *The Writing of Royalism, 1628–1660* (Cambridge: Cambridge University Press, 2001), pp. 11, 13.
68. Harvey, *De motu cordis et sanguinis*, pp. 3–4.
69. Andrew Lacey, "The Office for King Charles the Martyr in the *Book of Common Prayer*, 1662–1685," *Journal of Ecclesiastical History* 53 (2002): 510–14, 523–24.
70. Vergil, *Aeneid* 6.110–11.
71. See Waith, *Herculean Hero*, p. 11.
72. Galinsky, *Herakles Theme*, p. 6.
73. Rolf Soellner, "The Madness of Hercules and the Elizabethans," *Comparative Literature* 10 (1958): 309–24. Pseudo-Aristotle, *Problemata* 30.1 953a.
74. Erasmus, *De copia*, pp. 216, 60, 152, 180, 236; trans., pp. 591, 331, 481, 538, 611. See also the eloquent Gallic Hercules, ibid., pp. 206, 216; and Peacham, *Complete Gentleman*, p. 18. For the curriculum, see Leach, *Educational Charters and Documents*, p. 468.
75. Horace, *Odes* 3.3.9–12; trans. C. E. Bennett, *The Odes and Epodes*, rev. ed. (Cambridge, Mass.: Harvard University Press, 1927), p. 179.
76. Erasmus, *De ratione studii*, ed. Jean-Claude Margolin, in *Opera omnia* (Amsterdam), I-2:126; trans. Brian McGregor, in *Collected Works of Erasmus*, 24:676. *De ratione studii*, p. 131.

77. See Carl Bridenbaugh, *Vexed and Troubled Englishmen: 1590–1642* (New York: Oxford University Press, 1968), p. 43.
78. Erika Rummel, "The Reception of Erasmus' *Adages* in Sixteenth-Century England," *Renaissance and Reformation* 30 (1994): 19–25.
79. Baldwin, *William Shakespeare*, 2:749.
80. Harvey, *De motu cordis et sanguinis*, pp. 21, 59.
81. Erasmus, "Herculi labores," in *Adagia*, eds. M. L. van Poll-van de Lisdonk et al., in *Opera omnia* (Amsterdam), 2–5:24, 26, 28, 39. See also Galinsky, *Herakles Theme*, pp. 139–41.
82. George Chapman, *Chapman's Homer: The "Iliad," the "Odyssey" and the Lesser Homerica*, ed. Allardyce Nicoll, 2 vols. (New York: Pantheon for the Bolligen Foundation, 1956), 2:8, lines 203, 210.
83. See Galinsky, *Herakles* Theme, pp. 2–3, 102.
84. Vivian Nutton, *Ancient Medicine*, 2nd ed. (London: Routledge, 2012), pp. 51, 282, 280, 286, 35.
85. Marcus Aurelius, in Marcus Cornelius Fronto, *Epistulae*, ed. Michel J. van den Hout, 2 vols. (Pisa: Giardini, 1989), 1:27; cf. pp. 41, 75.
86. Lawrence J. Bliquez, *Roman Surgical Instruments and Other Minor Objects in the National Archeological Museum of Naples* (Mainz: Philipp von Zabern, 1994), pp. 99–106.
87. Jody R. Pinault, *Hippocratic Lives and Legends* (Leiden: E. J. Brill, 1992), pp. 7–10, 25, 35–60, 41, 126, 145, 147. See also François Ratief and Louise Cilliers, "The Epidemic of Athens, 430–426 B.C.," *South African Medical Journal* 88 (1998): 50–53.
88. See Charles F. Mullett, *The Bubonic Plague and England: An Essay in the History of Preventative Medicine* (Lexington: University of Kentucky Press, 1956), pp. 55, 64, 90, 125, 160, 164; Andrew Wear, *Knowledge and Practice in English Medicine, 1550–1650* (Cambridge: Cambridge University Press, 2000), pp. 231–32.
89. See Webster, "William Harvey," p. 3; Clark, *History of Royal College*, pp. 254–55.
90. *Prefaces to the First Editions of the Greek and Roman Classics and of the Sacred Scriptures*, ed. Beriah Botfield (London: H. G. Bohn, 1861), pp. 362, 364–65. Cf. Georg Agricola "with Hercules blessing." For *dextro Hercule*, see Erasmus, *Adagia*, in *Opera omnia* (Amsterdam), 2–1:186. See Jonathan Woolfson, *Padua and the Tudors: English Students in Italy, 1485–1603* (Toronto: University of Toronto Press, 1998), pp. 80–85. See also Nikolaus Mani, "Die griechische *editio princeps* des Galenos (1525), ihre Enstehung und ihre Wikung," *Gesnerus* 13 (1956): 29–52; Paul Potter, "The *editiones principes* of Galen and Hippocrates and Their Relation," in *Text and Tradition: Studies in Ancient Medicine and Its Transmission Presented to Jutta Kollesch*, eds. Klaus-Dieterich Fischer

et al. (Leiden: E. J. Brill, 1998), pp. 243–61. For the single detailed record of Linacre's medical practice, see Erasmus, *Epistolae*, eds. P. S. Allen et al., 12 vols. (Oxford: Clarendon, 1906–58), 6:46–47; trans. Alexander Dalzell, *Collected Works of Erasmus*, 11:68. See also Francis Madison, "Editor's Introduction," in *Linacre Studies: Essays on the Life and Work of Thomas Linacre, c. 1460–1524*, ed. idem et al. (Oxford: Clarendon, 1977), pp. xv–vi.

91. John A. Gee, *The Life and Works of Thomas Lupset: With a Critical Text of the Original Treatises and the Letters* (New Haven, Conn.: Yale University Press, 1928), pp. 59–63, 117–18.
92. Charles D. O'Malley, *English Medical Humanists: Thomas Linacre and John Caius* (Lawrence: University of Kansas Press, 1965), pp. 18–19.
93. Erasmus (trans.), Galen, *Exhortatio ad bonas artes praesertim medicinam, De optimo docendi genere, Et qualem oporteat esse medicum*, ed. Jan Hendrik Wasink, in *Opera omnia* (Amsterdam), 1–1:629–69.
94. See also Paul Oskar Kristeller, "The Humanist Movement," in idem, *Renaissance Thought and Its Sources*, ed. Michael Mooney (New York: Columbia University Press, 1979), pp. 21–32. For the term "medical humanism," see recently Nancy G. Siraisi, *History, Medicine, and the Traditions of Renaissance Learning* (Ann Arbor: University of Michigan Press, 2007), pp. 4, 261–62.
95. Erasmus, trans., Galen, pp. 631–35.
96. Erasmus, *Adagia*, in *Opera omnia* (Amsterdam), 2–5:29; trans. Margaret Mann Phillips et al., *Collected Works of Erasmus*, 34:172.
97. Harvey, *De motu cordis et sanguinis*, p. 20.
98. Boyle, "Harvey's Anatomy Book and Literary Culture"; "Reprising Terence's Plot"; "Harvey in the Sluice."
99. Charles Singer, "Some Vesalian Problems," *Bulletin of the History of Medicine*, 17 (1945): 426. See also pp. 427–29, 437–38. Johann Guinther von Andernach, *Institutionum anatomicarum secundum Galeni sententiam ad candidatos medicinae libri quatuor* (Basel: Balthasar Lasium and Thomas Platter, 1536), p. 46. Translation mine. For Massa's influence on Harvey, see French, *Harvey's Natural Philosophy*, pp. 22–26.
100. C. D. O'Malley, *Andreas Vesalius of Brussels* (Berkeley and Los Angeles: University of California Press, 1964), pp. 90–94, 46, without reference to *mehercule*.
101. Andreas Vesalius, *Anatomicarum Gabrielis Falloppii observationum examen*, in *Opera omnia anatomica et chirguria*, ed. Hermann Baerhaave et al., 2 vols. (Leiden: J. du Vivie and J. and H. Verbeek, 1725), 2:794.
102. Harvey, *De motu cordis et sanguinis*, pp. 20–21.
103. For invention, see Boyle, "Reprising Terence's Plot," and especially its definition on p. 371.

104. Ibid., p. 21.
105. Apollodorus, 2.5.5; Theocritus, *Idylls* 25; Pausanias, *Descriptio Graeciae* 5.1.10. For Theocritus, see Graham Zanker, "Pictorial Description as a Supplement for Narrative: Augeas' Stable in *Heracles Leontophonos*," *American Journal of Philology* 117 (1996): 411–23; Ivan M. Linforth, "Theocritus XXV," *Transactions of the American Philological Association* 78 (1947): 77–87. See also Pierre Brulé, "Héraclès et Augé: À propos d'origines rituelles du mythe," in *IIe recontre Héraclénne: Héraclès: Les femmes et le féminin*, eds. Colette Jourdain-Annequin et al. (Brussels: Institut historique belge de Rome, 1996), pp. 35–49. For the origin of the artistic cycle of his labors, see Bernard Ashmole and Nicholas Yalouris, *Olympia: The Sculptures of the Temple of Zeus* (London: Phaidon, 1967), pp. 23, 29, pls. 202–11; Frank Brommer, *Heracles: The Twelve Labors of the Hero in Ancient Art and Literature*, trans. and enlarged, Shirley J. Schwarz (New Rochelle, N.Y.: Aristide D. Caratzas, 1979), p. 30.
106. Nigel J. Spivey, *Songs on Bronze: The Greek Myths Made Real* (New York: Farrar, Strauss and Giroux, 2005), p. 55.
107. Coluccio Salutati, *De laboris Herculis*, ed. B. L. Ullman, 2 vols. in 1 (Zurich: In aedibus Thesauri mundi, 1951), pp. 365–67.
108. Jung, *Hercule*, p. 120. For a Hercules-Christ comparison in 1629, see John Milton's "Ode on the Morning of Christ's Nativity," cited by Galinsky, *Herakles Theme*, p. 205.
109. Erasmus, "Augeae stabulum repurgare," in *Adagia*, in *Opera omnia*, ed. Jean Leclerq, 11 vols. (Leiden, 1703–6), 2:530; trans., *Collected Works of Erasmus*, 33:201.
110. Leone Allacci, *Animadversiones in antiquitatum etruscarum fragmenta ab Inghiramo edita* (Paris: S. Cramoisy, 1640), pp. 1, 22, cited by Ingrid D. Rowland, *The Scarith of Scornello: A Tale of Renaissance Forgery* (Chicago: University of Chicago Press, 2004), pp. 74, 75.
111. Erasmus, *Epistolae*, 2:101. Erasmus, *Antibarbari*, ed. Kasimierz Kumaniecki, in *Opera omnia* (Amsterdam), 1–1:74–75, 59–60, 62–63; trans. Margaret Mann Phillips, *Collected Works of Erasmus*, 23:49, 25, 36–7.
112. Jacobus Sylvius, *Vesani cujusdam calumniarum in Hippocratis Galenique rem anatomicam depulsio* (Paris, 1551), fols. 4r, 28r, cited by O'Malley, *Vesalius*, pp. 248, 250, without reference to my argument. For Vesalius's impiety, see French, *Harvey's Natural Philosophy*, pp. 36–38. For Vesalius's praise of Sylvius, see his *De fabrica*, fol. 3r.
113. Harvey, *De motu cordis et sanguinis*, p. 18.
114. C. R. S. Harris, *The Heart and the Vascular System in Ancient Greek Medicine: From Alcmaeon to Galen* (Oxford: Clarendon, 1973), pp. 305, 313–14, 322, 333, 334; Walter Pagel, *William Harvey's Biological Ideas:*

Selected Aspects and Historical Background (Basel: S. Karger, 1967), pp. 129–30, 134, 156–69.

115. Harvey, *De motu cordis et sanguinis*, p. 12 for Matteo Realdo Columbo, *De re anatomica* (Venice: N. Bevilaqua, 1559).
116. Miguel Servet, *Restitutio christianismi* (Vienne, 1553).
117. Celsus, *De medicina* 4.1.4; trans. 1:357. Cf. 7.4.2. See also Jacques André, *La vocabulaire latin de l'anatomie* (Paris: Belles lettres, 1991), p. 139.
118. Columella, *Res rustica* 4.1.
119. Vergil, *Eclogae* 3.60, cited by Harvey, *Prelectiones anatomie universalis*, p. 2.
120. Vergil, *Eclogae* 1.34.
121. Harvey, *De motu cordis et sanguinis*, pp. 18, 32.
122. For contemporary English travelogues, see Thomas Coryat, *Coryat's Crudites*, 2 vols. (Glasgow: James Maclehose and Sons for the University, 1905), 1:270; Fynes Moryson, *An Itinerary*, 3 vols. (Glasgow: James Maclehose and Sons, 1907), 1:150. See also Vicenzo Dotto's map, *Pianta di Padova*, in Angelo Portenari, *Della felicità di Padova* (Padua: Tozzi, 1623). For Harvey and Paduan waterways, see also Boyle, "Harvey in the Sluice."
123. Moryson, *Itinerary*, p. 152.
124. Harvey, *De motu cordis et sanguinis*, pp. 32, 41.
125. See, e.g., French, *Harvey's Natural Philosophy*, pp. 90–93, 106, 372; Pagel, *Harvey's Biological Ideas*, pp. 24–25, 73–79; idem, "William Harvey Revisited (I) and (II)," *History of Science* 9 (1970): 1; Don G. Bates, "Harvey's Account of His 'Discovery,'" *Medical History* 36 (1992): 361–78, with *copia* as "large amount," p. 363.
126. Harvey, *De motu cordis et sanguinis*, pp. 41, 43, 44, 45, 46, 47, 48, 50, 52, 53, 54, 63, 72. For *quantitas*, see p. 44.
127. Leach, *Educational Documents and Charters*, p. 468.
128. Erasmus, *De copia*, pp. 34, 26, 34, 28, 26, 35–38, 280; trans., pp. 303, 295, 303, 298, 295, 658.
129. Harvey, *De motu cordis et sanguinis*, p. 41.
130. Ibid., pp. 5, 58.
131. Wolfgang Höfer, *Hercules medicus; sive, locorum communum liber* (Venice: Jacobus Kürner, 1657), frontispiece and preface.
132. Harvey, *De motu cordis et sanguinis*, p. 20. This comparison is not in Aristotle's extant works, although he cites Euripus for earthquakes and the purple murex in *Meteorologia* 2.8 366a; *Historia animalium* 5.15 547a.
133. Harvey, *De motu cordis et sanguinis*, p. 20, citing André DuLaurens, *Historia anatomica humani corporis*, in *Opera omnia anatomica et medica* (Frankfurt: William Fitzer, 1627), p. 352. If Harvey consulted this

edition, it would date his sentence, and it may have motivated the submission of his own manuscript to that publisher.

134. Jerome J. Bylebyl, "Disputation and Description in the Renaissance Pulse Controversy," in *The Medical Renaissance of the Sixteenth Century*, ed. Andrew Wear et al. (Cambridge: Cambridge University Press, 1985), pp. 226–27, 241–42.
135. Galen, *De usu partium libri XVII*, ed. George Helmreich, 2 vols. (Amsterdam: A. M. Hakkert, 1968), 1:331. Cf. *De methodo medendi*, in *Opera*, 10:649, 697. George Ent's *Apologia pro circuitione sanguinis* (Leiden: Peter Vander, 1687), which was dedicated to Harvey, included a formal digression on "the ebb and flow of the sea," pp. 64–100.
136. Procopius, *History of the Wars of Justinian* 8.6.20 10; trans. H. B. Dewing, *Procopius*, 7 vols. (Cambridge, Mass.: Harvard University Press, 1962–71), 5:107, 103–5, 107–9.
137. Pliny, *Naturalis historia* 2.99.218–2.100.219; Strabo, *Geographia* 1.2.30; Boethius, *De consolatione philosophiae* 2.1 meter. Cf. the tidal phenomenon of the straits of Gibraltar with the Pillars of Hercules, which the hero erected as his monument. Diodorus Siculus, *Bibliotheca* 4.18.4–5; Procopius, *Historia* 8.6.3, 8.6.8; Thomas Heywood, *The Brazen Age* (London: Nicholas Okes for Samuel Rand, 1613), pp. 38, 42. See also the title page of Bacon, *Instauratio magna*, 1620, and A. D. Burnett, *The Engraved Title-Page of Bacon's "Instauratio magna": An Icon and Paradigm of Science and its Wider Implications* (Durham: Thomas Harriot Seminar, 1998), p. 2 fig. 1, and pp. 6–11. For Bacon and Harvey, see also Keynes, *Life of Harvey*, pp. 157–61, 433; French, *Harvey's Natural Philosophy*, pp. 325–27. Cf. also the mouth of the Thames River, a native analogy that Harvey later explicates, inserting himself into the tradition of Aristotle's deliberation about Euripus with that English example. Ercole V. Ferrario and F. N. L. Paynter, "William Harvey's Debate with Casper Hofmann on the Circulation of the Blood: New Documentary Evidence," *Journal of the History of Medicine and Allied Sciences*, 15 (1960): 15. Hofmann seems not to have known that the Thames is a tidal river, which regularly reverses its flow. "Caspar Hofmann to William Harvey," Appendix I in Gweneth Whitteridge, *William Harvey and the Circulation of the Blood* (London: Macdonald, 1971), p. 242. For Hofmann, see French, *Harvey's Natural Philosophy*, pp. 255–58.
138. Nathanael Carpenter, *Geography Delineated Forth in Two Books*, 2nd ed. (Oxford: John Lichfield and William Turner for Henry Cripps, 1625), book 2, 82.
139. Cited by William T. Costello, *The Scholastic Curriculum at Early Seventeenth-Century Cambridge* (Cambridge, Mass.: Harvard University Press, 1958), p. 59.

140. Francesco Stelluti, "Al detto Sig. Galilei del sig. Francesco Stelluti, Accademico Linceo," in Galileo Galilei, *Le opere di Galileo Galilei*, eds. Antonio Favoro et al., rpt. of Edizione Nazionale, 20 vols. (Florence: S. A. G. Barbèra, 1929–39), 6:209; *The Controversy on the Comets of 1618*, trans. Stillman Drake et al. (Philadelphia: University of Pennsylvania Press, 1960), p. 159.
141. Erasmus, *De copia*, p. 104; translation mine. Erasmus, *Parabolae*, ed. Betty I. Knott, in *Opera omnia* (Amsterdam), 1–5:272. Erasmus, "Euripus homo," in *Adagia* (Amsterdam), 2–2:382–85; trans., *Collected Works of Erasmus*, 32:215–16.
142. Seneca, *Hercules Oetaeus* lines 774–841. For the heart and blood, see lines 1220–23; for the pillars of Hercules, line 1240.
143. Erasmus, *De conscribendis epistolis*, ed. Jean-Claude Margolin, in *Opera omnia* (Amsterdam), 1–2:337; trans. Charles Fantazzi, *Collected Works of Erasmus*, 24:88.
144. Heywood, *Brazen Age*, pp. 34–42, with citation p. 39. See also Shakespeare, *Antony and Cleopatra* IV.xii.43–47.
145. Harvey, *De motu cordis et sanguinis*, p. 20.
146. Ibid. Cf. Girolamo Fabrici d'Aquapendente, *De venarum ostiolis* (Padua, 1603), facsimile rpt., ed. Kenneth J. Franklin (London: Balliére, Tindall, and Cox, 1933), p. 72.
147. Ibid., pp. 72, 71, 75, 78, 80–81.
148. Erasmus, *De copia*, p. 208; trans., p. 282. The source is Xenophon, *Memorabilia* 2.1.21–34; but also Cicero, *De officiis* 1.32.118. See Erwin Panofsky, *Hercules am Scheidewege und andere antike Bildstoffe in der neuren Kunst* (Leipzig: B. G. Teubner, 1930); Waith, *Herculean Hero*, pp. 42–43; Emma Stafford, "Vice or Virtue? Herakles and the Art of Allegory," in *Herakles and Hercules: Exploring a Graeco-Roman Divinity*, ed. Louis Rawlings et al. (Swansea: The Classical Press of Wales, 2005), pp. 71–96.
149. Fabrici, *De venarum ostiolis*, pp. 75, 78, 80–81.
150. Seneca, *Hercules Oetaeus* line 1661.
151. Portenari, *Padova*, pp. 10, 11, 12.
152. Corrado Lattanzi, "L'attività giovanile de Bartlomeo Ammannati in Veneto," in *Bartolomeo Ammannati scultore e architteto, 1511–1592*, eds. Rosselli Del Turco et al. (Florence: Associazione dimore historiche italiane, Sezione Toscana: Alinea, 1995), pp. 88–93; Michael Kiene, *Bartolomeo Ammannati* (Milan: Electa, 1995), pp. 34–45; Lionello Puppi, "Il 'Colosso' del Mantova," in *Essays Presented to Myron Gilmore*, eds. Sergio Bertelli et al., 2 vols. (Florence: La Nuova Italia, 1978): 1:11–29; Joachim Poeschke, *Michelangelo and His World: Sculpture of the Italian Renaissance*, trans. Russell Stockman (New York: Harry N. Abrams, 1996), pp. 195–206.

153. Erasmus, "Herculanus nodus," in *Adagia* (Amsterdam), 2–2:367–8; "nodum solvere," 2–1:28, 118–9. For the knot of virginity (*hymen*) on the London stage, see Ben Johnson, "Hymnaei; or, The Solemnities of Masque and Barrier at a Marriage," in *Selected Masques*, ed. Stephen Orgel (New Haven, Conn.: Yale University Press, 1970), pp. 49, 342; William Shakespeare and George Willis, *Pericles* IV.ii.1146, cf. Shakespeare, *The Tempest* IV.i.16. See also Stephanos Panayotakis, "The Knot and the Hymen: A Reconsideration of *nodus virginitatis (Hist. Apoll.* 1)," *Mnemosyne* 53 (2000): 605, 603. See also Ann M. Nicgorski, "The Magic Knot of Herakles: The Propaganda of Alexander the Great, and Tomb II at Vegina," in *Herakes and Hercules*, pp. 97–102.
154. C. L. Day, *Quipus and Witches Knots: The Role of the Knot in Primitive and Ancient Cultures* (Lawrence: University of Kansas Press, 1967), pp. 53–57, citing Pliny, *Naturalis historia* 28.17.63–4. For Oribasius in translation, *De laqueis*, see pp. 117–18. For illustrations, see pp. 133, 138, 144.
155. Whitteridge, *William Harvey*, p. 83.
156. Harvey, *De motu cordis et sanguinis*, p. 20. Cf. DuLaurens, *Historia anatomica*, p. 352.
157. Erasmus, *Declamatio in laudem artis medicae*, ed. J. Domański, in *Opera omnia* (Amsterdam), I-4:164–66, 167, 168, 178–80, 184, 185–86; trans. Brian McGregor, *Collected Works of Erasmus*, 29:39, 45, 47–48, 49, 42.
158. Harvey, *Prelectiones anatomie universalis*, p. 175.
159. Keynes, *Life of Harvey*, pp. vii, 47.
160. Ernst R. Curtius, *European Literature and the Latin Middle Ages*, trans. Willard R. Trask (London: Routledge and Kegan Paul, 1953), pp. 132–33.
161. Elizabeth Spieler, *Science, Reading, and Renaissance Literature* (Cambridge: Cambridge University Press, 2004), pp. 71–72.
162. Thomas Fuller, *History of the Worthies of England* (London: J. G. W. L. and W. G. for Thomas Williams, 1662), cited by H. Bayon, "William Harvey, Physician and Biologist: His Precursors, Opponents, and Successors," *Annals of Science* 3 (1938): 85.
163. Edelstein, *Hippocratic "Oath,"* pp. 51–52.
164. Erasmus, *De copia*, p. 76; trans., p. 347. For cursing, see *De copia*, pp. 157–58.
165. Keynes, *Life of Harvey*, pp. 290, 316, 420, 397.
166. John Collop, "On Doctor Harvey," in *Poesis Redivia; or, Poesies Reviv'd*, in *The Poems of John Collop*, ed. Conrad Hilberry (Madison: University of Wisconsin Press, 1962), p. 102.

Index

M. O'R. Boyle, *Cultural Anatomies of the Heart in Aristotle, Augustine, Aquinas, Calvin, and Harvey*,
https://doi.org/10.1007/978-3-319-93653-6

Kolibri 360 GmbH
Ferdinand-Jühlke-Straße 7
99095 Erfurt, Deutschland
produktsicherheit@kolibri360.de